NANOCRYSTALLINE METALS AND OXIDES

Selected Properties and Applications

THE KLUWER INTERNATIONAL SERIES IN:

ELECTRONIC MATERIALS: SCIENCE AND TECHNOLOGY

Series Editor

HARRY L. TULLER
Massachusetts Institute of Technology

Other Books in Series

DIAMOND: Electronic Properties and Applications
Lawrence S. Pan and Don R. Kania, Editors
ION IMPLANTATION: Basics to Device Fabrication
Emanuel Rimini, Author
SOLID STATE BATTERIES: Materials Design and Optimization
C. Julien and G. Nazri, Authors
SOL-GEL OPTICS: Processing and Applications
L.C. Klein, Editor
PHOTOREFRACTIVE EFFECTS AND MATERIALS
David Nolte, Editor
MATERIALS FOR OPTOELECTRONICS
Maurice Quillec, Editor
PIEZOELECTRIC ACTUATORS AND ULTRASONIC MOTORS
Ken Uchino, Author
WIDE-GAP LUMINESCENT MATERIALS: Theory and Applications
Stanley R. Rotman, Editor
THIN FILM FERROELECTRIC MATERIALS AND DEVICES
R. Ramesh, Editor
MICRO-ACTUATORS: Electrical, Thermal, Optical, Magnetic, Mechanical and Acoustic
Massood Tabib-Azar, Author
HIGH-TEMPERATURE SUPERCONDUCTORS: Materials, Properties and Applications
Rainer Wesche, Author

NANOCRYSTALLINE METALS AND OXIDES

Selected Properties and Applications

edited by

Philippe Knauth
Professor
Université de Provence
Marseille, France

Joop Schoonman
Professor
Delft University of Technology
Delft, The Netherlands

KLUWER ACADEMIC PUBLISHERS
Boston / Dordrecht / London

Distributors for North, Central and South America:
Kluwer Academic Publishers
101 Philip Drive
Assinippi Park
Norwell, Massachusetts 02061 USA
Telephone (781) 871-6600
Fax (781) 871-6528
E-Mail < kluwer@wkap.com>

Distributors for all other countries:
Kluwer Academic Publishers Group
Distribution Centre
Post Office Box 322
3300 AH Dordrecht, THE NETHERLANDS
Telephone 31 78 6392 392
Fax 31 78 6546 474
E-Mail < services@wkap.nl>

Electronic Services < http://www.wkap.nl>

Library of Congress Cataloging-in-Publication Data

Nanocrystalline metals and oxides: selected properties and applications/edited by Philippe Knauth, Joop Schoonman.
p. cm—(Electronic materials – science and technology)
ISBN 0-7923-7627-7 (alk. Paper)
1. Nanostructure materials. 2. Crystals. 3. Electronic ceramics. I. Knauth, Philippe II. Schoonman Joop. III. Kluwer international series in—electronic materials, science and technology.

TA418.9.N35 N327 2002
621.381'04—dc21

2001053973

Printed on acid-free paper.

CONTENTS

PREFACE

Nanostructured materials have at least one dimension in the nanometer range. They became a very active research area in solid state physics and chemistry in recent years with anticipated applications in various domains, including solar cells, electronics, batteries and sensors.

Nanocrystalline metals and oxides are dense polycrystalline solids with a mean grain size below 100 nm. This book is intended to give an overview on selected properties and applications of nanocrystalline metals and oxides by leading experts in the field. In such a rapidly growing and evolving domain, the selection of properties to be covered is always somewhat arbitrary and reflects the personal bias of the editors.

The first three chapters provide a very complete theoretical treatment of thermodynamics and atom/ion transport for nanocrystalline materials.

J. Weissmüller gives an overview on the thermodynamics of materials with small size or a large interface density, including a discussion of capillary parameters, single-component materials and alloys, stress and strain effects. The main focus is on metallic and covalent solids. The chapter includes the detailed thermodynamic formalism necessary for the discussion of size effects.

J. Bernardini and D. L. Beke review diffusion in nanomaterials. Diffusion theory in presence of stationary and moving grain boundaries is discussed with consideration of segregation and stress effects. The analysis is then extended to the nanoscale, where the classical Fick laws are not always applicable. Diffusion on a discrete lattice, in amorphous materials and reaction/nucleation problems are treated.

J. Maier discusses thermodynamics and transport in the bulk and at boundaries of ionic materials. Point defects and ionic conductivities in the bulk as well as core and space charge effects at interfaces are analyzed. Size effects on defect chemistry and the domain of nanosized ionic and mixed conductors are explored. The chapter ends with an outlook on nanoionics and soft materials science.

The following chapters are expert's views on the development of experimental characterization techniques for nanocrystalline solids with emphasis on electroceramic materials. Two chapters are devoted to specific applications of nanocrystalline oxides, i.e., gas sensing and photoelectrochemical cells, where the potential advantages over microcrystalline samples are particularly evident.

T. O. Mason and co-workers show how impedance spectroscopy can be used by skilled and careful experimentalists to separate effective grain boundary and grain interior resistances and capacitances of nanocrystalline samples. The authors suggest also modifications of the usual brick-layer model in order to analyze nanoscale ceramics.

A. V. Chadwick and G. E. Rush introduce EXAFS spectroscopy as a powerful tool to investigate the defect structure of dense nanocrystalline oxides. The main conclusion of their careful analysis is the picture of a nanocrystalline oxide as well-ordered crystallites separated by grain boundaries similar to those of conventional microcrystalline samples.

M. I. Baraton shows how FTIR spectroscopy can be applied to study surface layers of nanocrystalline oxides. The variation of transmitted infrared energy can be related to electrical conductivity changes. The reactions of nanocrystalline oxides with gases can be monitored "in situ" under the real working conditions of gas sensors.

E. Traversa et al. present gas sensing properties of nanocrystalline oxides, used in potentiometric and resistive sensors. The basic mechanisms of these devices are shortly described. Advantages and drawbacks of various nanocrystalline oxides, including SnO_2 and TiO_2, are discussed and directions for future experiments are suggested.

Finally, A. Zaban discusses basic concepts of the photo-electrochemistry of nanoporous electrodes, especially the construction and operation of dye-sensitized solar cells based on nanosize TiO_2. The special features of nanoporous electrodes, such as electrical potential distribution, electron motion, interface energetics, are outlined.

The book is intended for a broad range of readers, foremost chemists, physicists, and materials scientists. Theoretical physicists and chemists will certainly also profit from this book. The electroceramics and solid state ionics community are particularly addressed, given the main interests of the editors.

We thank all the colleagues who spend considerable time and effort writing these high level contributions. We are also pleased to acknowledge the support of the series editor Prof. Harry Tuller and of Greg Franklin, senior editor at Kluwer Academic Publishers.

Finally, P. K. gratefully acknowledges the support by the French Centre National de la Recherche Scientifique (CNRS) and the National Science Foundation (NSF) that helped to realize this and other projects on nanocrystalline metals and oxides.

J. S. acknowledges the European Science Foundation-NANO program and the Delft Interfaculty Research Center "Renewable Energy" for support of exchange visits and nanoparticle research.

Marseille, France, and Delft, The Netherlands, September 2001.

P. Knauth and J. Schoonman

NANOCRYSTALLINE METALS AND OXIDES

Selected Properties and Applications

THERMODYNAMICS OF NANOCRYSTALLINE SOLIDS

Jörg Weissmüller
Forschungszentrum Karlsruhe, Institut für Nanotechnologie, Germany
and
Universität des Saarlandes, Technische Physik, Germany
Joerg.Weissmueller@int.fzk.de

1. INTRODUCTION

Thermodynamics have provided the phenomenological framework for understanding phase equilibria and the reaction of materials to changes in temperature and pressure for several centuries. It was recognized early on that surfaces play an important role for the equilibrium state, as is evidenced by the pressure-increase in small droplets, analyzed quite generally by Young and Laplace two hundred years ago. Important results from the 19th century, such as the Gibbs-Thomson-Freundlich equation for the increased vapor pressure of a droplet, Ostwalds theory of coarsening of dispersions, and Gibbs comprehensive treatment of capillarity, emphasize the importance of interfacial phenomena. The effects of capillarity are most pronounced in small systems or, more generally, systems with a large interfacial area per volume, and the long known results of the thermodynamics of interfaces have therefore remained of importance as experimental studies in the recent decades took an enhanced interest in materials with a nanometer-scale structure.[1] However, not all of the intriguing properties of such materials are adequately described by the conventional theory; this is due to the fact that many 'nanomaterials' are solids and, although Gibbs discussed some aspects of the thermodynamics of solids, most of the results at the time were devised with attention to fluids. The generalization of the formalism to solids is among the achievements in thermodynamics in the 20th century. Specifically, the notion of surface or interface stress, established by Shuttleworth, and the continuum mechanics of open systems due to Larché and Cahn prove to be of relevance in the context of nanocrystalline materials.

This article discusses issues related to the thermodynamics of materials with a small size or a large number of interfaces, with specific attention to solids. The focus is on nanocrystalline materials, that is, dense polycrystalline solids with a nanometer-scale grain size, where the interfaces are grain boundaries, rather than free surfaces or heterophase interfaces. One of the issues which will be addressed concerns the differences between these materials and isolated particles.

Some synthesis routes for nanocrystalline materials involve processes which may be analyzed in terms of dynamic equilibria in driven systems, and an understanding of the evolution of the microstructure upon annealing is the subject of theories for nonequilibrium processes, such as grain growth. Driven systems and nonequilibrium processes are ignored in the present contribution, which is focused instead on the impact of having a large number of grain boundaries on such states of the material for which meaningful information can be derived from equilibrium thermodynamics. Emphasis is on effects which are intrinsically related to the presence of interfaces, whereas extrinsic defects, which are often a consequence of the highly nonequilibrium processing routes of nanocrystalline materials, are disregarded.

Detailed discussion on structure and energetics of internal interfaces can be found in Refs. 2 and 3, and more general treatments of capillarity are given, for instance, in Refs. 4, 5 and 6. An earlier treatise of the thermodynamics of nanocrystalline materials was given in Ref. 7.

2. REMARKS ON THE STRUCTURE OF GRAIN BOUNDARIES IN NANOCRYSTALLINE SOLIDS

Because knowing the atomic structure of grain boundaries is of interest for understanding their thermodynamic properties we shall start by considering the present understanding of this issue, which is reviewed in Refs. 8, 9 and 10.

Much of the early interest in the field of nanocrystalline materials was stimulated by reports[11,12] of a state of matter in the grain boundaries with a considerably lower degree of atomic short-range order (SRO) than other solid states of matter. Indeed, further experimental evidence has indicated that grain boundaries in nanocrystalline metals, in particular immediately after preparation, can exhibit states with enhanced energy compared to conventional grain boundaries, or with more disordered grain boundary cores.[13,14,15] However, these states are found to relax to states of lower energy or higher order, and the more recent results of experiment[15,16,17,18] and of computer simulation[19,20,21,22] provide little support for the assumption that the grain boundary structure of relaxed nanocrystalline materials differs from that of high-angle boundaries in macroscopic bicrystals. Previous reports to that extent can apparently be largely qualified as artefacts related to impurities at grain boundaries. The present understanding is that, in nanocrystalline as well as coarse-grained metals, the atomic short-range order in the core region of the grain boundaries differs from that of the perfect crystal lattice with respect to the coordination number, local symmetry, or bond-length distribution, and that this region appears to be quasi two-dimensional, with a width of essentially one interatomic distance. Computer simulations of covalently bonded materials suggest wider disordered grain boundary regions than in metals,[21] but there appears to be as yet no systematic experimental study of high-angle boundaries in covalent nanocrystalline solids which might corroborate this prediction.

3. CAPILLARY PARAMETERS

In the regions immediately adjacent to surfaces and interfaces the materials properties are different from the bulk properties. Exploring the atomic-scale properties by theory requires elaborate numerical computations, either ab initio or by molecular dynamics; these are available in exceptional cases only, and few such results have been verified by comparison to experiment because measuring local properties of internal interfaces on an atomic scale is an equally demanding task. By contrast, phenomenological thermodynamics describes relations between quantities which are defined for macroscopic systems or for smaller regions which can be approximately described by extrapolating the equations of state of macroscopic systems; such relations are more readily combined with available experimental data. Different formal approaches may be preferred, depending on the problem being analyzed, and it is instructive to shortly address a selection of the more important ones.

3.1 Specific Excess

Most often, the experimentally accessible information is limited to the *total* value of the extensive parameters, and it is then useful to analyze the effect of interfaces in terms of the total excess of the actual property over an appropriate reference state of the material with negligible interfacial area. We shall briefly reproduce the derivation for the relation of the specific interface free energy γ to the excess ΔG in the Gibbs free energy. Let γ be defined in terms of the work required to form an element of surface,

$$dE = T\,dS - P\,dV + \sum_j \mu_j\,dN_j + \gamma\,dA. \qquad (1)$$

From the definition of the Gibbs free energy, $G = E - TS + PV$, it then follows by Legendre transform that

$$\gamma = \partial G/\partial A\,|_{T,P,N_j}. \qquad (2)$$

By applying Eulers theorem to Eq. (1) it is also found that

$$E = TS - PV + \sum_j \mu_j N_j + \gamma A\,, \qquad (3)$$

which implies

$$G = \sum_j \mu_j N_j + \gamma A. \qquad (4)$$

The sum represents the free energy of the matter in the system, if it was reversibly inserted into a macroscopic system with negligible interface area and with the same chemical potentials as the actual material.[6] This state of the matter in the system is the reference state corresponding to the present choice of state variables (P,T,N_j), and we denote its free energy by the subscript 'R': $G_R = \sum_j \mu_j N_j$. Note that this reference state is hypothetical and that, in general, it will not correspond to an equilibrium state of the matter in the system. It is seen that, on top of being the derivative of G with respect to the area, Eq. (2), γ is also the excess of the free energy in the actual system over the free energy in the reference state, per area:

$$\gamma = \left(G - G_R\right)/A. \qquad (5)$$

The term in brackets denotes the total excess, ΔG, of the free energy.

The constitutive assumption $E = E(S,V,N_j,A)$, which underlies Eq. (1), is only adequate when the excess energy depends on the presence of interfaces exclusively through A. When the energy depends, in addition, on the grain size, the interface curvature, or the separation between interfaces, then it may be required to verify carefully whether the capillary energy defined by Eq. (2), the derivative of a free energy with respect to the area, is identical to the one obtained from Eq. (4) as the total excess per area.

In coarse-grained materials the properties of the bulk phases sufficiently far from an interface are not affected by the presence of the interface. In this case only the interfacial regions contribute to the specific excess, in other words, the specific excess is representative of the *local* properties of the matter at the interfaces. This is illustrated in Fig. 1a), which shows schematically the variation of a local free energy density across an interface; the area of the hatched region represents the specific excess. But when the interfaces are only nanometers apart, then it is possible that the properties of the matter in the interior of the bulk phases are significantly modified due to the presence of the interfaces. Several examples for this effect will be discussed in this article: the increased free energy density in the bulk arising from interface-induced long-range stresses (Section 6.2-4), the stress-induced change in solute concentration in the bulk for open systems (Section 7.1), and the depletion of solute in the bulk of a closed system due to the partitioning of solute between interfaces and bulk (Section 5.1-2). In consequence, the actual values of the densities in the bulk of nanoscale materials will differ from the corresponding density in the reference system; this implies that the excess quantities are no longer representative of the local state of matter at the interfaces, but that they represent the added contribution of interfaces *and* bulk to the total values of the extensive variables, as illustrated schematically in Fig. 1b).

A further difference between macroscopic and nanoscale microstructures

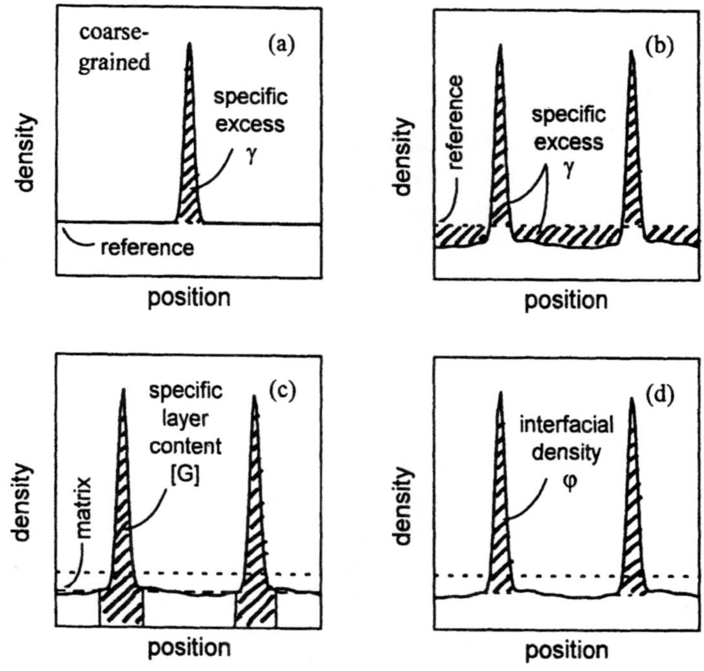

Figure 1. Schematic diagram of the variation of the free energy density as a function of the position across a single grain boundary in a coarse-grained material (**a**) and across two neighboring grain boundaries in a nanocrystalline material (**b** - **d**). The area of the hatched regions represents the various interfacial quantities; in (**b**) the contribution of the bulk to γ is negative. Full lines: actual free energy density; dotted lines: value for the reference state; dashed line in (**c**): value for the matrix.

resides in the sensitivity of the value of A to the precise way in which the dividing surface, $\mathcal{S}$, which defines the position of the interface, is located. For plane surfaces, both A and γ are independent of the precise location of $\mathcal{S}$ (p. 234 in Ref. 4) but the same is not true when the surface is curved or when there are edges, as exemplified in Sections 4.2-3 below. It is therefore important to have as precise as possible a notion of the location of $\mathcal{S}$. In fluids, the Young-Laplace equation provides the means for defining uniquely the position of the 'surface of tension' (p. 229 in Ref. 4), but for reasons which are addressed in Section 6 below this method fails in solids. The constitutive assumption $E = E(S,V,N_j,A)$ requires that γ is independent of the curvature, and in some instances this requirement provides a criterion for a unique location of $\mathcal{S}$ in a solid particle, see Section 4.2 below. It may also be argued that, since structural probes suggests very narrow grain boundary cores (compare Section 2), it will generally be perfectly adequate to use Gibbs' notion (p. 219 in Ref. 4) of having the dividing surface 'coincide sensibly with the physical surface of discontinuity' (i.e., the core of the boundary) in conjunction with the structural information to locate $\mathcal{S}$ in grain boundaries.

3.2 Specific Layer Content

The excess formalism is useful in analyzing experimental data which lack information on the detailed spatial variation of the materials properties, but many modern experimental techniques do discriminate between the state of the matter at interfaces on the one hand and the matter in the bulk on the other, see, for instance, Section 7.1-2. Moreover, simple microscopic models for interfaces will generally work with explicit assumptions for separate equations of state, or for specific atomic interactions, of the interfacial regions and the bulk. Rather than computing excess quantities based on such data or models, thereby ignoring the available microscopic insights, it is then often more appropriate to work with the 'layer formalism',[6] which allows to decompose the thermodynamic functions into contributions from interfacial layers of finite thickness and from the regions of bulk phase (the 'matrix') outside the layers, irrespective of the detailed spatial variation of the properties across the interfaces. The layer thickness and the location of the layer bounds can be arbitrarily chosen, but it is most useful to have the layer contain all the interfacial regions with modified properties. Of course, the layer bounds are not physical interfaces; instead, they represent the - up to a certain degree arbitrary - convention on how to subdivide the system into interfacial and bulk regions.

The quantities corresponding to the specific excess are here the specific layer contents, defined as the total value of the extensive quantities in the layers divided by the area; they are denoted by square brackets. For instance, the total free energy is

$$G = G_M + A\,[G]. \tag{6}$$

where G_M is the total free energy in the matrix. Figure 1c) illustrates the definition of the specific layer content, and shows that a major difference to the excess formalism is that the matrix models the matter in the bulk of the nanocrystalline system in its *actual* state, whereas the specific excess is computed relative to the *reference* state of a macroscopic system with negligible interfacial area.

The layer formalism provides an alternative formulation of capillary phenomena which is equivalent to the analysis in terms of excess quantities, but which has the advantage of being more directly related to microscopic models which provide separate equations of state for interfacial regions and the bulk. It is possible to derive all the result of thermodynamics in terms of the layer properties, and to convert layer quantities to Gibbs excess and vice versa,[6,23] but only such combinations of layer quantities which are independent of the location of the layer bounds carry physical meaning. For a comprehensive discussion of the reader is referred to Ref. 6.

When implementing atomistic models or specific equations of state, such as the regular solution approximation, it can be convenient to work with a layer model in which the material is divided into subsystems containing fixed amounts of material;[24] when $P\,V$ terms in the free energy are negligible then the position of the layer bounds can be ignored. The subsystems represent, respectively, the matter in the bulk and in the interfacial regions, and their properties are modeled by independent equations of state. For the example of a binary alloy, the respective molar free energy functions are $g_M(T, x_M)$ and $g_L(T, x_L)$, where x is a solute fraction and the subscripts refer to matrix and layer. The effective molar free energy of the system (total free energy over total amount of matter) is then a weighted average over the subsystems,

$$\bar{g} = (1 - \nu)\; g_M(T,x_M) + \nu\; g_L(T,x_L), \tag{7}$$

with ν the fraction of matter (solvent plus solute) in the layers. The grain size enters this approach through the parameter ν, which is proportional to the grain boundary area. At equilibrium, subject to the constraint of constant ν (constant grain size), the diffusion potentials $\partial g_M/\partial x_M$ and $\partial g_L/\partial x_L$ must be equal. This allows to eliminate the internal parameters, the local solute fractions, and to write the effective molar free energy of the closed system as a function of the macroscopic state variable, the overall solute fraction:

$\bar{g} = \bar{g}(T,\bar{x},\nu)$. For sufficiently large values of ν such models predict that the resulting free energy function $\bar{g}$ may depend on the state variables in a qualitatively different way compared to the original functions g_M and g_L - see Section 5 and Refs. 25, 26, 27.

3.3 Interfacial Density

Some problems require even more spatial information than the mere separation into interfacial layers and bulk. The most obvious example is provided by continuum mechanics, where the stress **S** is a function of the position **x** which can be obtained as the derivative of a nonuniform (Helmholtz) free energy density $\Phi(T,\mathbf{E})$ in the bulk with respect to the strain **E**. The phenomena of wetting[28] and of spinodal decomposition[29] in fluids can only be understood in terms of extra energy due to gradients of the concentration,[30] and consequently their description also requires the consideration of a nonuniform free energy density. By a suitable coarse-graining of atomistic details one can define local densities, per volume, for each extensive thermodynamic variable (for instance, concentration, free energy density, or density of matter), and treat them as continuous functions of the position. Discontinuous interfaces can be incorporated in such approaches by introducing *interfacial densities*, per area,[28,31] for instance an interfacial free energy density φ. For example, the total Helmholtz free energy of a homogeneous solid phase in this notation is

$$H = \int \Phi(T,\mathbf{E})\, \mathrm{d}v + \int \varphi(T,\mathbb{E})\, \mathrm{d}a\,, \tag{8}$$

where the integration is over the total volume and total interfacial area, respectively, and where $\mathbb{E}$ refers to the tangential strain, see Section 6.1 below. Equation (8) implies that φ is a local excess, at interfaces, over the bulk in its *actual* state, as illustrated in Fig. 1d). An example where fundamentally different results are obtained by analysis of the interfacial free energy density φ as opposed to the specific excess free energy γ will be given in the definition of the so-called 'interface stress' in Section 6.6.

4. SINGLE-COMPONENT MATERIALS

4.1 Intrinsic instability; grain growth

Similar to homogeneously disordered solids, such as glasses, nanostructured solids have a higher free energy than the equilibrium state, the single crystal. The excess free energy varies essentially as the inverse of the grain size; at small size it is quite large, comparable to the enthalpy of fusion.

In order to understand the stability of nanocrystalline materials it is therefore important to understand and measure the thermodynamic functions.

While nanocrystalline solids have an excess free energy comparable to that of glasses, the two types of systems differ in so far as glasses are *metastable* with respect to formation of the equilibrium phase, that is with respect to crystallization, whereas poly- or nanocrystalline materials are *unstable* with respect to the approach to equilibrium by grain growth. There is a long history to the question whether one can fill space by cells (bubbles in a foam, or grains in a polycrystal) in such a way that the interface area is minimum for a given mean cell size; this would imply that idealized polycrystals with a suitable arrangement of the grain boundaries could actually be metastable. Kelvin had proposed a periodic array of tetrakaidecahedra ('Kelvin bodies') as a candidate for a minimum area tessellation of space,[32] and it is only quite recently that a configuration with lower area was found with the help of modern computer programs.[33] The same technique helped finally to demonstrate that all known configurations are unstable with respect to a transformation where the lattice constant is doubled and half the cells shrink whereas the other half grows, a typical instance of grain growth;[34] therefore, polycrystals with a positive grain boundary excess free energy are always unstable.

Experimentally it is found that pure nanocrystalline substances undergo grain growth even at temperatures which are a small fraction of the melting temperature T_m.[35,36] In the case of Pd, the grain size was observed to double within a few weeks of ageing *at room temperature*, corresponding to a homologous temperature T/T_m as low as 0.16.[36] This demonstrates the need to explore concepts for stabilizing the grain size, for instance by manipulation of the energy or mobility of grain boundaries.

4.2 Size- or curvature-dependent interface free energy.

Since the interfacial excess free energy γ contributes so strongly to the total free energy of nanocrystalline materials it is of interest to know its magnitude, and in this context a possible size-dependence of γ has been the subject of discussion. An experimental study by calorimetry reported that for nanocrystalline (nc-) Se the total excess of the enthalpy, ΔH, increases less steeply than $1/D$ with diminishing grain size D, and it was concluded that the specific grain boundary enthalpy decreases with decreasing grain size.[37] Molecular dynamics computations for fcc metals by several authors arrive at the same finding for the size-dependence of ΔH,[19,38,39] and they provide an indication for the microscopic origin of the deviation from the $1/D$ law. In fact, one of the most intriguing findings of the computer simulations is that, microscopically, the excess enthalpy of the grain boundaries is localized in an interfacial layer of constant width t, essentially independent of the grain size and wider than the structural core of the boundary,[20,21,39] and that triple lines and higher order grain junctions are just intersections of layers of constant t and constant free energy density.[39,62] It is instructive to analyze this situation for the example of an isolated nanoparticle.

In terms of the simulation result, the total free energy for an isolated, spherical nanocrystallite of radius r and with a surface layer of increased free energy density is

$$G = g_R \rho_R \frac{4\pi}{3}(r-t)^3 + g_L \rho_L \frac{4\pi}{3}\left[r^3 - (r-t)^3\right] \tag{9}$$

where g and ρ denote the molar free energy and molar density, and the subscripts 'R' and 'L' refer to the reference system and to the surface layer, respectively. By computing the specific surface free energy as total excess per area of the outer surface of the particle one finds that, to first order in r/t,

$$\gamma \approx \gamma_0 \,(1 - t/r). \tag{10}$$

The term $\gamma_0 = (g_L \rho_L - g_R \rho_R)\, t$ denotes the specific excess free energy of a plane surface, and the term which depends on r implies a reduction of γ when the surface is curved. Obviously, this nontrivial prediction is a simple geometric consequence of the concept of an interfacial layer of constant thickness and free energy density. Assuming t=0.35 nm (corresponding to a grain boundary layer thickness of twice that value, 0.7 nm, as suggested by the molecular dynamics results for fcc metals) and a grain size D = 5 nm for the purpose of illustration, it is found that γ is reduced, relative to γ_0, by 14 %.

The finding that the total excess free energy per area varies with r is of interest for predicting the thermodynamic functions of small particles, but it is incompatible with the standard treatments of capillarity based on the constitutive assumption that the total energy E is a function of the state variables S, V, N, and A alone. This follows because, when γ depends on r, then one can conceive sets of particles with a distribution of sizes which have identical values of S, V, N, A, but different E depending on the functional form of the distribution. A simple solution to this problem is to locate the dividing surface, not in the outer surface of the particle, but at a radius which makes γ independent of r. Strictly speaking, this is the only location of the dividing surface which is consistent with the formula in Section 3.1. For the present example, an excess per area which is independent of r is obtained when the radius of the dividing surface is $r - t/2$.

4.3 Triple lines

In contrast to the constant surface curvature of the spherical particle, the mean curvature of grain boundaries is generally less than $2/r$; instead, the boundaries are more or less planar and their orientation changes discontinuously at triple lines, that is, lines where three grain boundaries meet. It is then more adequate to attribute deviations from the $1/D$ variation of ΔH to the variation of the total length of triple lines and to a specific excess free energy λ per length of the lines, rather than to a dependency of γ on the surface curvature.[38] The total energy is then assumed to obey $E = E(S, V, N, A, L)$, and λ is defined so that

$$dE = T\,dS - P\,dV + \sum_j \mu_j\,dN_j + \gamma\,dA + \lambda\,dL. \tag{11}$$

By analogous arguments as in Section 3.1 it is then shown that λ is an excess, per length of edge or triple line, of the solids total free energy over that of a reference system with identical S, V, N, and A. Simple dimensional arguments[40] show that the length of linear elements, per volume, will increase as $1/D^2$ with decreasing grain size, so that at very small size the linear elements become relatively more important compared to the surfaces, which have the weaker size-dependence $A \propto 1/D$.

When λ is defined as a specific excess, then the molecular dynamic simulations find that at at least some of the triple line must have a negative value for λ.[38] It is also possible to generalize the layer formalism to include volume elements along the grain boundary junctions which represent the lines, and it was found that, contrary to λ, the enthalpy per length of these elements is positive.[39]

For truly one-dimensional elements of the microstructure, negative values of λ are excluded by a stability criterion:[41] for negative λ the polycrystal is unstable since it could indefinitely reduce its free energy by creating more and more triple lines at constant grain boundary area. The fact that this process is not observed in the computer models seems to suggest that the stability criterion cannot be transferred to lines and surfaces afflicted with a volume.

4.4 Size-Dependent Chemical Potential: Isolated Small Particle versus Nanocrystalline Solid

It is well known that the chemical potentials μ_j of the constituents (labeled by j) of isolated small particles depend systematically on the grain size, and one might assume that a similar dependency holds for nanocrystalline solids.

However, as we shall see, this is indeed *not* the case, and in order to illustrate this we shall briefly recapitulate the isolated particle case.

As above, we can decompose the total free energy of the particle as the sum of the free energies of reference system and excess. In this case,

$$G(T,P,N,A) = N\,\mu_R(T,P) + \gamma A \tag{12}$$

with $\mu_R(T,P) = \partial G_R(T,P,N) / \partial N$ an equation of state of the reference phase. For a given N the form of the particle which minimizes the free energy is obtained by the Wulff construction,[42] which determines a fixed geometrical shape at equilibrium. Since the shape is given, the area, at equilibrium, is uniquely defined by the volume or, more precisely, by the number of atoms:

$$A = c\,(\Omega N)^{2/3} \tag{13}$$

with c a dimensionless constant which depends on the particle shape and Ω the atomic volume. This has the important consequence that A and N cannot be varied independently, and that Eq. (12) can be reduced to

$$G(T,P,N) = N\,\mu_R(T,P) + \gamma\, c\,(\Omega N)^{2/3}. \tag{14}$$

By definition, the chemical potential of the particle is then

$$\mu(T,P,N) = \frac{\partial G}{\partial N}\Big|_{T,P} = \mu_R(T,P) + \frac{2}{3}\gamma\, c\,\Omega^{2/3} N^{-1/3}. \tag{15}$$

This is a general form of the Gibbs-Thomson-Freundlich equation. For the special case of a spherical particle, where $\Omega N = 4\pi r^3/3$, it takes on the more familiar form

$$\mu = \mu_R(T,P) + 2\gamma\Omega / r. \tag{16}$$

Among the ramifications of Eq. (16) is the well-known increase in vapor pressure at small particle size, and a change of phase equilibria in multicomponent systems since the individual constituents suffer different increase in μ due to their different partial atomic volumes.

When trying to derive an equivalent result for polycrystals or condensed nanocrystalline solids, one is immediately confronted with the fundamental difference that, for given N, the equilibrium value of the grain boundary area A is simply zero. Of course, this corresponds to the single crystal case which is not of interest in the present context. But in a poly- or nanocrystalline material, where A is finite, it is always possible to vary A independently of N, so that a fundamental requirement in the derivation of Eqs. (15) and (16) is violated. Therefore, the Gibbs-Thomson-Freundlich equation does not apply to nanocrystalline solids.

Note that, if the definition of μ was to be followed rigorously, then since $G = G(T, P, N, A)$ in a polycrystal, and since all variables are independent, we would have $\mu = (\partial G/\partial N)|_{T,P,A}$. This is a variation at constant grain boundary area and, in as much as the specific excess γ is a constant, the value of μ thus obtained would be independent of γ, in striking contrast to Eqs. (15) and (16). However, this consideration appears to be of little relevance, since it is not obvious how the variation of N at constant A could be realized experimentally for a nanocrystalline material, at least when it is an elemental solid: experiments which probe the chemical potential of single-component materials require that processes of dissolution and accretion are active, processes which will generally allow some sort of irreversible coarsening to occur simultaneously with the measurement, thereby changing A in an essentially uncontrolled way. These complications make the precise discussion of the meaning of experimental data for the electromotoric force of these materials a challenging task.[43,44]

The situation can be more favorable in the case of alloys. As we shall see in the next Section, certain alloys may exhibit metastable equilibrium states for which the grain boundary area A is finite and a function of the remaining thermodynamic state variables. Varying the amount of matter at equilibrium will then result in a defined variation of the area; in other words, the chemical potentials are then defined in terms of a variation which can be realized in an experiment. Yet another instance where the chemical potential can be measured in a straightforward way is an interstitial solid solution where only the solute is mobile, whereas the solvent atoms are frozen in such a way that the host lattice is constrained to purely elastic deformation. The solute fraction can then be varied at equilibrium while the host lattice is maintained. This situation will be discussed in Section 7 below.

5 NANOCRYSTALLINE ALLOYS

5.1 Metastable nanocrystalline alloys with finite grain size and vanishing γ

It has been proposed that, as a consequence of the Gibbs adsorption equation, alloys with a large enthalpy of grain boundary segregation may exhibit a reduced or even vanishing Gibbs excess free energy of the grain boundaries.[23,]

[25,45] Such alloys would then be metastable, since grain growth will not reduce the total Gibbs free energy unless solute is withdrawn from the alloy by nucleation and growth of the equilibrium phases. What follows is an essentially qualitative discussion of the concept.

In a binary solid solution polycrystal, solute atoms can be located on lattice sites and on grain boundary segregation sites; in many alloys the latter are energetically preferred. The heat released when a small amount of solute is transferred from the lattice to the segregation sites, per mol of transferred solute, is called the enthalpy of segregation, ΔH^{seg}. Consider the limit of small grain size D (large grain boundary area A), so that there are more segregation sites than solute atoms. At T=0, and if ΔH^{seg} is positive, the free energy is minimized when all solute is segregated, and the solid can further reduce its Gibbs free energy by reducing the total grain boundary area A through grain-growth. Grain growth reduces the number of segregation sites, and leads to a state (with area A^0 and grain size D^0) where there are as many such sites as solute atoms. If the grain size grows beyond D^0, then solute originally located in the grain boundary must be re-dissolved in the crystal lattice. Hence, on decreasing A below A^0, interface energy is gained, but enthalpy of solution must be provided to transfer solute from the energetically favorable lattice sites to the unfavorable grain boundary sites. The variation of the total Gibbs free energy for this process is

$$\gamma = \partial G/\partial A\,|_{P,T,N_1,N_2} = \gamma_0 - \{N_2\}\,\Delta H^{seg} \tag{17}$$

at $T = 0$ and $A < A^0$, where γ_0 is the specific excess free energy in the pure solvent, N_1, N_2 denote the total amount of solvent and solute, respectively and $\{N_2\}$ is the specific excess of solute. It is seen that γ may be of either positive or negative sign, depending on the magnitude of ΔH^{seg}. If γ changes sign at $D = D^0$, then any further variation of A will increase the free energy, in other words the solid is in a stable or metastable state at the finite grain size $D = D^0$. Unless new phases of lower free energy nucleate, no grain growth will take place with increasing temperature, until the entropy of mixing favors the gradual dissolution of solute atoms in the crystals. More quantitative investigations of the problem can be found in the literature, including the effect of entropy at $T > 0$.[23,25] Figure 2 a) displays an example for the variation of G with D for large ΔH^{seg}; the minimum in G is here at $D^0 \approx 15$ nm, and the corresponding change of sign in γ is seen in Fig. 2 b).

The large ΔH^{seg} required to make the right-hand side of Eq. (17) negative presupposes a large and positive heat of solution in the crystal lattice. Such alloys have very small equilibrium solubilities, and it is therefore believed that grain size stabilization can only be achieved in supersaturated solutions.[23,25]

This is further emphasized by Gibbs statement (p. 240 in Ref. 4) that γ must be positive in stable phases. The nanocrystalline alloy will then be metastable, as opposed to stable, and this state can be maintained as long as the nucleation of the stable phase is hindered by kinetic barriers.

It may be objected that the idea of the topological lattice defect grain boundary having a negative specific energy is against physical intuition. However, since γ is the *total* variation of the solids Gibbs free energy with the grain boundary area (compare Section 3.1), a negative value of γ does not imply that the specific 'topological defect energy' which is associated with a grain boundary (due to e.g. broken bonds or non-dense atomic packing, to mechanical strain fields in the crystalline matrix, and to a change in chemical short-range order for the atoms situated in the interface) is negative. Instead, it is implied that the *combined* effect of forming a new segment of grain boundary, where additional solute atoms can be incorporated, and transferring the solute atoms from their energetically unfavorable positions in the lattice to the more favorable positions in the grain boundary, is an overall decrease of the solids Gibbs free energy.

The concept of grain size stabilization by segregation has seen some experimental support, both in the form of a systematic variation of the grain size with the solute content in metals[46] and ionic crystals,[47] in agreement with theoretical predictions and, more directly, by the observation that significant grain growth in nc-Pd can be delayed to temperatures as high as 1000 °C by the addition of Zr, an element with a large enthalpy of segregation.[26] In solids it is still a matter of debate whether these findings may possibly be due to kinetic hindering of grain growth by solute drag,[48] but there is a whole class of

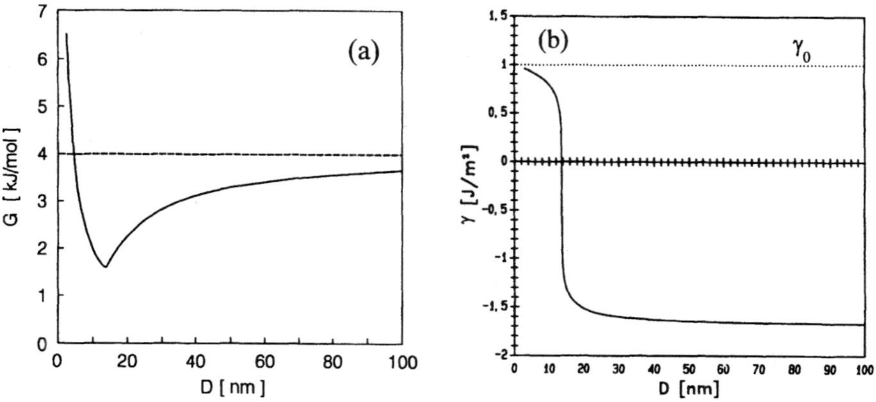

Figure 2. Variation of the Gibbs free energy *G* (**a**) and of the specific excess free energy γ (**b**) of grain boundaries in a nanocrystalline alloy with a large enthalpy of segregation as a function of the grain size *D*. Overall solute fraction is 0.05, and temperature *T* = 600 K. See Ref. 23 for further details.

fluid substances, the microemulsions, for which the true thermodynamic stabilization of nanometer-scale droplets at vanishing interfacial energy is well established.[49,50]

5.2 Destabilization of compounds by grain boundary segregation

Interestingly, the Pd-Zr alloy system, which exemplifies the stabilization of nanocrystalline solid solutions by segregation, has provided an example that segregation or, more generally, the partitioning of solute between comparable number of sites at the grain boundaries and in the crystal lattice, may also have the opposite effect, namely to destabilize the nanocrystalline material: stoichiometric Pd_3Zr, a congruently melting intermetallic line compound, can be prepared in the nanocrystalline state by high energy ball milling. Upon annealing it was found that the nanocrystalline material decomposes into a mixture of the Pd-Zr solid solution and the compound. The decomposition could be explained in terms of the partitioning of Zr between grain boundaries and the bulk, which destabilizes the compound by depleting it in Zr.[51] Indeed, it is found that the material transforms back to the single-phase equilibrium state as the grains grow to a macroscopic size, and the number of segregation

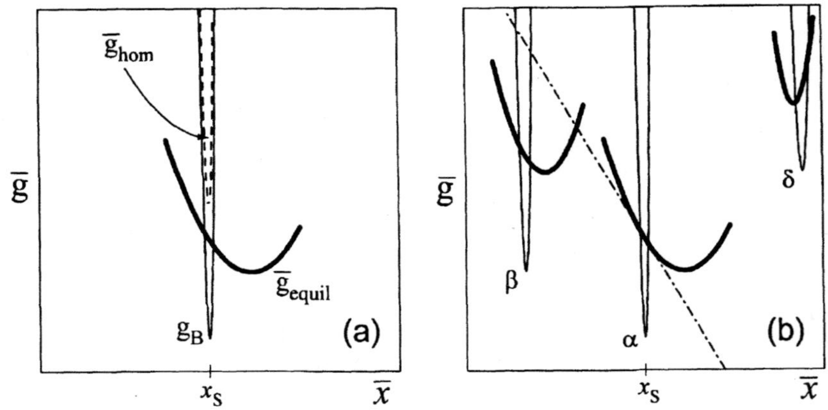

Figure 3. Schematic molar free-energy functions g for nanocrystalline compounds. (**a**): As compared to the molar free energy of the bulk phase, g_B, the curve for the homogeneous nanocrystalline compound $\bar{g}_{hom}$ (no grain boundary segregation) simply has a constant free energy $\gamma_0 A$ added. Equilibration of grain boundaries and bulk leads to segregation and, thereby, to the curve $\bar{g}_{equil}$, which is much shallower and has a displaced minimum. (**b**): Schematic free energy diagram of three coarse-grained (thin solid lines) and nanocrystalline (bold solid lines) phases α,β,δ. Nominally stoichiometric (solute fraction x_S) nanocrystalline phase α is unstable with respect to precipitation of β since the free energy of β is below the tangent to α at x_S (dashed line). See Ref. 26 for details.

sites is progressively reduced. A schematic illustration of the free energy diagram for nanocrystalline compounds is reproduced in Fig. 3.

6. STRESS AND STRAIN AT INTERNAL INTERFACES

For isolated solid nanoparticles one generally finds that the lattice constant depends measurably on the particle size. This is a consequence of the surface stress, a fundamental thermodynamic quantity which has become accessible to experiments only comparatively recently. The growing body of data on the stress of the single crystal-vacuum surface is reviewed in Ref. 52. There are also several studies of interfaces in metallic multilayers (see reviews in Refs. 53 and 54), but the interface stress associated with grain boundaries has not been systematically explored. As we shall see, the interface stress can drastically change the thermodynamic equilibrium, in particular in solid solutions, and it is therefore of interest to examine the issue of stress and deformation of grain boundaries, and the relevant equilibrium equations.

6.1 Deformation of grain boundaries

The grain boundary free energy varies as a function of the relative crystallographic orientation of the abutting crystal lattices (the 'macroscopic degrees of freedom') and of the relative displacements of the two crystals tangential and normal to the interface ('microscopic degrees of freedom').[2] Similar to bulk phases, where the stress can be defined as the derivative of a free energy with respect to the strain, the derivatives of γ or φ with respect to these configurational parameters define forces which act at the boundary and which need to be balanced by stresses in the bulk. Conversely, strain of the bulk phases changes the state of the grain boundary and, thereby, its free energy. In order to explore the mechanical equilibrium at an interface it is not useful to combine a bulk free energy function depending on the strain with an interfacial free energy function depending on crystallographic angles; instead, the continuum mechanics of internal interfaces can be formulated more conveniently when the configurational degrees of freedom for deforming the interface are expressed in terms of the displacements and displacement gradients in the bulk phases. The first approach in this direction is due to Shuttleworth,[55] who described the deformation of a solid free surface in terms of a 2×2 tensor in the tangent plane, and defined an 'interface stress' tensor as the conjugate quantity.

Similar to free surfaces, grain boundaries can be modeled as two-dimensional objects, and it might therefore seem that their continuum mechanics are simpler than those of bulk phases which, as three-dimensional objects, would be expected to have more degrees of freedom for deformation than interfaces. However, upon closer inspection, it is realized that the energetics and, therefore, the stress and strain, of internal interfaces depend actually on a quite considerably larger number of variables than that of bulk phases: first, even though interfaces are two-dimensional objects, the mechanical equilibrium between the interface and the bulk depends on the orientation of the interface normal in three-dimensional space, and on the interface curvature. Second, the interfacial energy depends on the strain in *both* phases which abut at the interface, that is the energy and, consequently, the generalized stresses are functions of *two* strain tensors, as opposed to the single tensor which describes the strain in the bulk. Thirdly, whereas the mechanical equilibrium in the bulk can often be analyzed within the framework of small displacement gradients

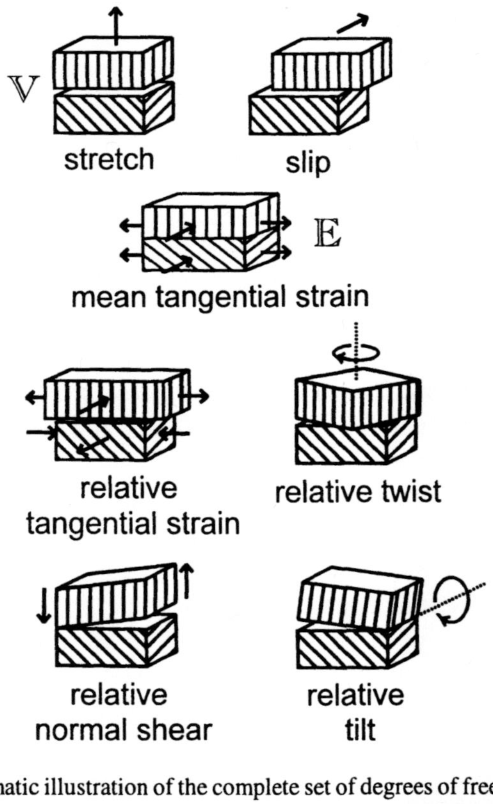

Figure 4. Schematic illustration of the complete set of degrees of freedom for deformation of internal interfaces. The individual parameters can be represented in terms of vectors and tensors defined on the interface. Generalized interface stresses, which are conjugate to the individual degrees of freedom, can be related to the stresses in the bulk phases via local balance equations.[56]

(small strain), stress at interfaces will generally result in finite *jumps* in the displacement vector, either in the form of relative slip of the abutting phases tangential to the interface, or in the form of stretch, that is a change of the interfacial excess volume. In spite of its complexity, the problem of deformation and of mechanical equilibrium at internal interfaces is open to a general analysis. Based on the postulate of frame invariance of the free energy, the most general set of degrees of freedom for interfacial deformation can be identified; the deformation is described in terms of the limiting values of the displacement field in the bulk, and of its gradients, at the interface. Figure 4 illustrates these quantities schematically. Conjugate to the vectors and tensors which represent the deformation are generalized interface stresses; these can be linked to the stresses in the bulk phases abutting at the (generally curved) interface by local balance equations, and elastic constants for the interface can be defined in terms of derivatives of the generalized interface stresses with respect to the strains.[56]

Only two of the degrees of freedom for interfacial deformation change the volume, namely the mean tangential strain, $\mathbb{E}$, and the normal jump in the displacement, or stretch, $\mathbb{V}$ (see Fig. 4). Because volume changes can be measured with comparative ease, the stretch and the conjugate quantity to $\mathbb{E}$, the interface stress $\mathbb{S}$, are also the only parameters in this context for which there are experimental data. We shall now discuss $\mathbb{S}$; further comments on $\mathbb{V}$ can be found in Section 7.2.

6.2 Interface stress; capillary equation for solids

The interface stress $\mathbb{S}$ is defined in terms of the work required to stretch the surface elastically.[55] This statement is commonly written in the form $\mathbb{S} = A^{-1}\, \mathrm{d}(\gamma A)/\mathrm{d}\mathbb{E}$ [where $\mathbb{E}$ denotes the mean tangential strain, see Fig. (4)], but we shall argue below (Section 6.6) that a more appropriate definition must be based on the surface free energy density, φ, rather than the specific excess γ; $\mathbb{S}$ is then defined through $\mathbb{S} = A^{-1}\, \mathrm{d}(\varphi A)/\mathrm{d}\mathbb{E}$. When the interfacial area is measured in laboratory coordinates, then this implies

$$\mathbb{S} = \varphi\, \mathbb{U} + \partial\varphi / \partial\mathbb{E}\,, \tag{8}$$

with $\mathbb{U}$ a tangential unit tensor on the surface.[a] The term $\varphi\mathbb{U}$ accounts for the

---

[a] Equation (18) is commonly written in terms of 2×2 tensors defined in the plane of the interface. This is appropriate for plane surfaces, but when the surface is curved or when several surfaces of different orientation contribute simultaneously to the balance of force in a three-

fact that the mean tangential strain $\mathbb{E}$ changes the interfacial area, thereby increasing the total free energy by $\varphi A\,\mathrm{trace}(\mathbb{E})$. In general, it is more convenient to measure the interfacial area in coordinates of the undeformed lattice (Lagrangian coordinates); the area is then constant, independent of $\mathbb{E}$, and one obtains the less familiar, but much simpler formula[6]

$$\mathbb{S} = \partial\varphi / \partial\mathbb{E} . \tag{19}$$

The difference between Eqs. (18) and (19) is analogous to that between the Cauchy and Piola-Kirchhoff stress tensors in three-dimensional bodies. Both definitions are used in the literature, and it is important to state that they lead to the identical numerical value for $\mathbb{S}$; they merely differ with respect to the functional dependency of φ on $\mathbb{E}$. On top of its formal simplicity, Eq. (19) also has the added advantage to illustrate quite unambiguously that the entries of the interface stress tensor $\mathbb{S}$ are independent thermodynamic quantities which are unrelated to the value of φ by magnitude and sign.

$\mathbb{S}$ can be experimentally measured by the stresses and strains in the bulk phases which are required to balance $\mathbb{S}$. The appropriate balance law is readily derived in terms of a virtual work argument. For an arbitrary solid microstructure, loaded by a uniform external pressure P_{ext}, the integral of the bulk stress $\mathbf{S}$ over the volume and the integral of the interface stress $\mathbb{S}$ over the total area of external surface and internal interfaces are related by the balance equation[57]

$$\int (\mathbf{S} + P_{ext}\, \mathbf{U})\, \mathrm{d}v + \int \mathbb{S}\, \mathrm{d}a = 0 . \tag{20}$$

$\mathbf{U}$ denotes the unit tensor. Particularly useful is the trace of this equation, which relates the volumetric mean of the pressure in the bulk to the mean of a scalar interface stress, $f = ½\, \mathrm{trace}\, \mathbb{S}$, on the surface and internal interfaces:

$$3\, V \langle P - P_{ext} \rangle_V = 2\, A \langle f \rangle_A . \tag{21}$$

$\langle\rangle_V$ and $\langle\rangle_A$ denote the volumetric and areal mean, respectively. When the elastic constants for the bulk phases are known, then the mean stresses for use with Eqs. (20) and (21) can be determined from experimental strains, for instance computed from lattice constant changes or from substrate bending. For the example of isotropic strain the pressure is related to the change in the

---

dimensional body, then a more general notion of interface stress is required. This is provided by the concept of superficial tensors, which allow to represent $\mathbb{S}$ and $\mathbb{E}$ as 3×3 tensors containing the additional information on the orientation of the surface in three-dimensional space.[31,56]

lattice constant a relative to a reference state at vanishing mean stress by $P = 3K\varepsilon$ with $\varepsilon = \Delta a/a$ the linear strain and K the bulk modulus; the strain is then given by

$$\langle \varepsilon \rangle_V = -\frac{2A}{9KV} \langle f \rangle_A . \tag{22}$$

Let us further assume an idealized polycrystal with spherical grains, for which the interfacial area per volume depends on the grain size, D, by $A/V = 3/D$. In this case, Eq. (22) has the simple form[57]

$$\langle \frac{\Delta a}{a} \rangle_V = -\frac{2}{3KD} \langle f \rangle_A . \tag{23}$$

Apparently there have been few systematic attempts at investigating the interface stress of grain boundaries experimentally. Hoffmann and Birringer[58] investigated the variation of the lattice constant of inert-gas condensed nc-Pd as a function of the grain size. These authors point out that when the x-ray extinction length is less than the sample thickness then the experimental lattice constant is not representative of the volumetric mean strain which is required to compute the mean pressure for use with Eq. (21). However, they show that the mean pressure can still be computed when the residual stress is corrected for. By computing the interface-induced stress from the known value of the bulk modulus in combination with the lattice strain relative to the coarse-grained material they deduced $f = +1.3(3)$ N/m. For the same material, Weissmüller and Lemier obtained the smaller value $f = +0.1(4)$ N/m.[59] The sample in the latter experiment had previously been exposed to hydrogen, and the difference may be due to irreversibly trapped hydrogen at grain boundaries; this would agree with the tendency of f to take on negative values when grain boundaries are deliberately enriched in hydrogen (compare Section 7.2 below).

6.3 Failure of the Young-Laplace equation in solids

The interface-induced stress in nanocrystalline solids has an analogy in the capillary pressure in fluid droplets, which obeys the Young-Laplace equation: across a curved interface between two fluids, the pressure jumps by

$$\Delta P = \gamma \left(1/r_1 + 1/r_2\right) \tag{24}$$

with r_1 and r_2 the principal radii of curvature. In fact, this equation is still frequently used to explain size-dependent changes of the lattice constant in solid nanoparticles. There are several reasons why this is wrong, and why the generalized capillary equation for solids, Eq. (20), must be preferred. In fluids, the pressure is a constant in each phase, and the mean curvature is therefore a constant along each surface. By contrast, the curvature of solid surfaces can vary in an arbitrary way as a function of the position on the surface; it is then not obvious which value of the curvature to use with Eq. (24). What is more, solid surfaces are often faceted, which implies that the curvature is not even systematically related to the grain or particle size. Therefore, the use of the integral quantities, total surface area and total volume, rather than the local curvature, is appropriate in solids. In fact, Eq. (24) is not even locally correct in solids, first because the appropriate capillary parameter is the tensor $\mathbb{S}$, not the scalar γ, and second because the local balance law for the jump $\Delta\mathbf{S}$ in the bulk stress across a solid interface,[31,56]

$$\Delta\mathbf{S}\;\mathbf{n} = -\operatorname{div}_S \mathbb{S}\,, \tag{25}$$

where div_S denotes the surface divergence, does not determine the jump in all the components of the stress tensor, nor does it determine the jump in the pressure. The mean values of stress and strain in Eqs. (20-22) are also the appropriate parameters in conjunction with experimental data which, most frequently, represent averages of the stress or strain in the bulk, or of the displacement on the surface, rather than local values of the pressure jump.

6.4 Size-dependent interface free energy due to interface-induced strain

Since the interface-induced stress can be quite large at small grain size, the associated strain and the mechanical work performed by the interface stress might possibly contribute in a significant way to the total excess free energy. This work is intrinsically related to the presence of the interfaces, and its value, per area of interface, must therefore be counted as a part of the interfacial excess, even though the extra strain energy is not entirely localized at the interfaces, but contributed partly by the bulk. In relation to the discussion of a size-dependent interfacial free energy in Section 4.2, it is of interest to estimate its magnitude.

We consider a spherical particle of diameter D, and we use, as the reference state for the strain, the stress-free particle; the excess free energy in this state is $\gamma_0 A$ with γ_0 the specific free energy of the undeformed surface. Let us compute the work W_A and W_V done by the induced strain against f on the surface and against the induced pressure P in the bulk, respectively, when the particles relaxes to the configuration of mechanical equilibrium; for simplicity, we shall neglect entropy effects and identify the work with a change in the free energy. The action of the strain ε in response to f is such that the specific free energy of the surface will be decreased at the expense of an increased free energy density of the bulk. The work done against the stress in the bulk is simply

$$W_V = \frac{9}{2} K \varepsilon^2 V, \tag{26}$$

and the change in the surface or interface energy is

$$W_A = 2 f A \varepsilon. \tag{27}$$

By using Eq. (22) for the interface-induced pressure strain we obtain for the total strain energies W_A and W_V:

$$W_V = \frac{2}{9} \frac{1}{K} \frac{A^2}{V} f^2 , \qquad W_A = -\frac{4}{9} \frac{1}{K} \frac{A^2}{V} f^2 . \tag{28}$$

Adding both contributions, and computing the excess per area using $A/V = 3 / D$ for grain boundaries supplies[10]

$$\Delta\gamma_f = -\frac{2A}{9KV} f^2. \tag{29}$$

Note that here, as in many other computations of the thermodynamic properties of nanomaterials, the grain size dependence differs, depending on whether nanocrystalline solids or isolated free particles are considered; in the present case idealized *isolated* spherical particles have a specific surface area $A/V = 6/D$, twice the value of the grain boundary area in an idealized *nanocrystalline* material of the same grain size. Consequently, the prefactor on the right-hand side of Eq. (29) must be doubled for isolated particles. For nanocrystalline Pd, with[58] $f = 1.3$ N/m and $K = 187$ GPa, Eq. (29) implies $\Delta\gamma_f = -0.0012$ J/m^2 at $D = 5$ nm, a negligibly small change. However, f may be larger for isolated particles, and the doubled surface to volume ratio may further increase the effect; thus, the surface stress may possibly have a measurable effect on the excess free energy of small isolated particles.

6.5 Lagrangian versus Laboratory Coordinates

The excess free energy in Eq. (29) is measured per area of the undeformed surface (since the derivation is based on Lagrangian coordinates), and it turns out to be interesting to compute the excess per area of the strained surface, $\hat{\gamma}$, for comparison - we recall that this is the more commonly used convention. In laboratory coordinates the area changes in proportion to the strain, and when contributions to the size-dependence other than the mechanical relaxation are ignored then $\hat{\gamma} = (\gamma_0 + \Delta\gamma_f)/(1 + 2\langle\Delta a/a\rangle)$ where γ_0 refers to the undeformed surface. With Eqs. (22) for the strain and (29) for $\Delta\gamma_f$ this yields[60]

$$\hat{\gamma} = \gamma_0 \left[1 - 2f^2/\left(3\gamma_0 KD\right)\right]/\left[1 - 4f/\left(3KD\right)\right]. \qquad (30)$$

for nanocrystalline materials (the result for isolated particles is obtained by substituting *D/2* for *D*).

According to Eq. (29) the excess per referential area is always a decreasing function of $1/D$. By contrast, an inspection of Eq. (30) shows that the excess per area of the deformed surface $\hat{\gamma}$ can be either a decreasing or an increasing function of $1/D$, depending on the sign of the term $f - 2\gamma_0$. This is due to the change in the area: for instance, when $f=2\gamma_0$ then the strain reduces the excess free energy in proportion to the area, so that their ratio $\hat{\gamma}$ is a constant, independent of D. The different size-dependencies of the functions $\gamma(D)$ and $\hat{\gamma}(D)$ illustrates once again (compare also Section 4.2) the importance of having a precise notion of how the interfacial area is defined and measured.

6.6 Defining Equation for the Interface Stress

We shall now reconsider the commonly employed definition of the interface stress in terms of a derivative of the specific excess free energy γ with respect to the strain of the surface. To this end, we compute the total excess per area for the strained particle by adding the work terms, Eqs. (26) and (27), to the excess, $\gamma_0 A$, of the undeformed particle, and we determine f tentatively according to $f = \partial\gamma/\partial\varepsilon$. The result is

$$f = \frac{\partial\gamma}{\partial\varepsilon} = \frac{\partial}{\partial\varepsilon}\left(\gamma_0 + 2f\varepsilon + \frac{9}{2}K\varepsilon^2\, V/A\right) = 2f + 9K\varepsilon\, V/A.$$

If this equation was evaluated at equilibrium, where the strain is given by Eq. (22), then it would be found that $f = 0$. This is not surprising since the condition for equilibrium is that the free energy is at minimum, and consequently its derivative with respect to the strain must vanish. Of course, this does not imply that there is no interface stress in the relaxed particle - its presence is clearly evidenced and measured by the strain. Instead, the implication is that the definition of an interface stress in terms of the change of γ as a function of the strain is not a useful one. Note that this statement is not limited to small particles - $f = 0$ would be found for *any* value of A/V, even in macroscopic bodies. The finding illustrates that the concept of interface stress refers to local forces acting at the interface, and that it requires, therefore, to be cast in a definition which involves a knowledge of the repartition of the energy between the interfacial regions and the bulk. Therefore, the formulation (Section 6.2) of $\mathbb{S}$ and f in terms the derivative of the surface free energy density, φ, with respect to the strain must be preferred over the conventional one in terms of the total excess per area, γ.

7. INTERACTION OF INTERFACE-INDUCED STRESS WITH COMPOSITION

7.1 Effect of the interface stress on the lattice constant in a solid solution

In the previous section we have seen that the interface stress in elemental nanocrystalline solids can be measured by analysis of the size-dependence of the lattice constant arising from interface-induced stresses. In solid solutions, the lattice constant is also a function of the solute fraction, and measured lattice constants are frequently the basis for inferring the solute concentration in the crystal lattice of alloys. In nanocrystalline solid solutions both effects may be present simultaneously, and it is then not immediately obvious if and how the elastic strain arising from the interface-induced stress can be separated from the composition-induced change in the lattice constant. Clearly, neglecting any one of the two contributions may lead to erroneous results for the other.

In spite of the added complication, the simultaneous chemical and mechanical equilibrium can be rigorously analyzed, and stress and solute fraction can be determined independently, when the lattice constant data are supplemented by measured values of the solute chemical potential μ.[59] The volume of a solid depends on its composition, and there is consequently also an interrelation between the conjugate state variables, the pressure and the chemical potential. In the simplest case the equation of state for μ is of the form

$$\mu(T, P, x) = \hat{\mu}(T, x) + 3\eta P / \rho, \tag{31}$$

where $\hat{\mu}$ denotes the chemical potential at $P = 0$, η is a concentration-strain coefficient, $\eta = (3V)^{-1}\,\partial V/\partial x|_{T,P}$, and the solute fraction x is defined so that the concentration is ρx, with ρ the density of sites available for solute. Conversely, the equilibrium value of the concentration at a given value of μ depends on the pressure, and a change in the pressure at constant μ will be accompanied by two types of strain: an immediate elastic response to the stress, and a further change of the lattice constant due to the change of the equilibrium solute fraction by diffusion. The first rigorous analysis of the relevant equilibrium conditions is due to Larché and Cahn,[61] who showed that the results of continuum mechanics are immediately transferrable from the conventional situation of constant composition to that of constant chemical potential, provided that the constant-composition elastic constants are replaced by the so-called 'open system elastic constants'. For instance, the open system bulk modulus, $K^* = -V^{-1}\,\partial V/\partial P|_{T,\mu}$, is related to the bulk modulus at constant composition, K, by

$$K^*(T,P,\mu) = K(T,P,\mu) \,/\, \left[1 + 9\,\eta^2 K(T,P,\mu)\,\chi(T,P,\mu)\right]. \qquad (32)$$

χ denotes the solute susceptibility, $\chi = \rho^{-1}\,\partial x/\partial\mu|_{T,P}$.

Constant-composition elastic constants are often only moderately dependent on the temperature, the concentration and - for not too large strains - on the pressure; therefore K may be approximated as a constant, and conventional problems can be well approximated within the framework of linear elasticity. By contrast, K^* cannot normally be assumed constant, since the solute concentration depends nonlinearly on μ, so that χ varies considerably with P. It follows that the strain in open systems must generally be treated as a nonlinear function of the stress. For the example of a dilute solution, where $\hat{\mu} = \mu_0 + RT\ln(x)$, we have $\chi = \hat{x}\exp[-3\eta P/(\rho RT)]/(\rho RT)$ where $\hat{x}$ denotes the equilibrium solute fraction at $P = 0$, and R denotes the gas constant. The stress-strain relation for the change in volume, relative to that of the stress-free alloy at the given value of μ, is then

$$-\frac{\Delta V}{V} = \int_0^P \frac{dP}{K^*(T,P,\mu)}$$
$$= \frac{P}{K} + 3\eta\,\exp\!\left(\frac{\mu - \mu_0}{RT}\right)\left[1 - \exp\!\left(-\frac{3\eta P}{\rho RT}\right)\right]. \qquad (33)$$

Similar equations can be derived for concentrated solutions, and the stress-strain relation thus found can be combined with experimental lattice constant data to derive the pressure; the interface stress is then computed by inserting the pressure into Eq. (21), and the concentration is found by solving Eq. (31) for x. This procedure is particularly useful when the concentration can be varied reversibly by equilibrating the nanocrystalline sample with a reservoir of solute at an adjustable chemical potential. Metal-hydrides provide the possibility to do just that, with the added incentive that the thermodynamic properties are highly relevant for their performance as hydrogen storage materials.

Figure 5a) displays experimental data[59] for the variation of the lattice constant in nc-Pd-H as a function of the hydrogen partial pressure p_{H_2}, which determines the hydrogen chemical potential through $\mu_H = \frac{1}{2} RT \ln p_{H_2}$. It is seen that, in comparison to single-crystalline Pd, the lattice constant of nc-Pd-H varies differently with p_{H_2}. This suggests that interface-induced stresses act on the crystal lattice of nc-Pd-H, and the analysis outlined in the previous section has been used by the authors of Ref. 59 to quantify the interface induced pressure and the interface stress. The results, Fig. 6 a), show that the interface stress in the Pd-H α-phase is negative, inducing a negative pressure of up to -0.5 GPa which expands the crystal lattice. A positive interface stress, and a positive interface-induced pressure of about +1.7 GPa was found for the same sample in the hydrogen-rich α'-phase.[27]

Figure 5b) shows the hydrogen fraction x_V in the bulk of nc-Pd-H as a function of p_{H_2}. It is instructive to compare this data to different other measures for the concentration, which are also displayed in the figure:[59] the solute fraction in single-crystalline Pd is seen to be significantly lower than x_V, a finding which emphasizes the influence of the interface-induced pressure on the equilibrium concentration. Clearly, however, most of the large increase in

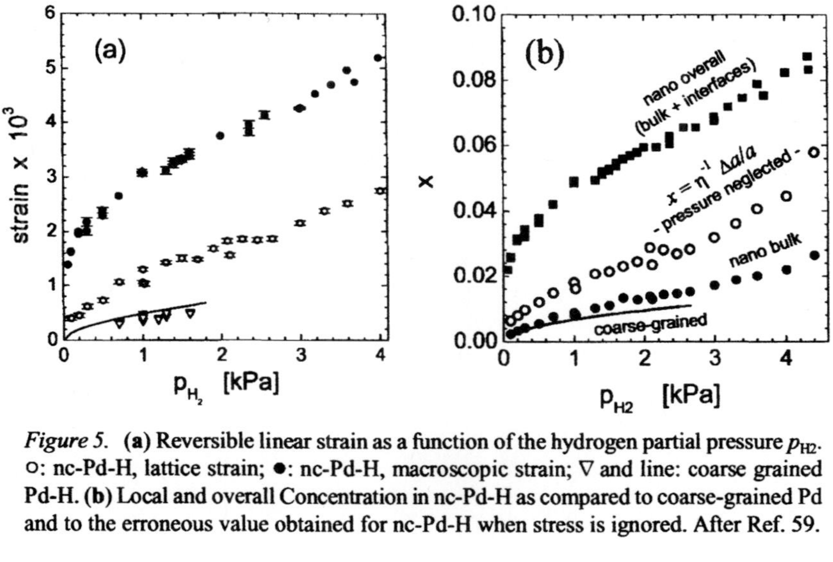

Figure 5. (**a**) Reversible linear strain as a function of the hydrogen partial pressure p_{H2}. o: nc-Pd-H, lattice strain; •: nc-Pd-H, macroscopic strain; ∇ and line: coarse grained Pd-H. (**b**) Local and overall Concentration in nc-Pd-H as compared to coarse-grained Pd and to the erroneous value obtained for nc-Pd-H when stress is ignored. After Ref. 59.

the total hydrogen fraction (overall hydrogen per metal atom) arises from enrichment of the hydrogen at grain boundaries, an effect which is readily understood in terms of the conventional theory of grain boundary segregation, and which had been pointed out earlier.[62] Finally, Fig. 5b) also displays the value of x_V which would be obtained if the solute fraction was computed from the lattice constant data based on the Vegard-type law $\Delta a/a = \eta\, x_V$, ignoring the interface-induced pressure. It is eminent that this procedure, which is routinely used in the literature, provides values for x_V in nc-Pd-H which are so seriously erroneous as to provide almost no useful information on any of the true concentration values. This implies that, at small grain size, the consideration of the interface-induced pressure is indispensable for a valid determination of the solute fraction from lattice constant data.

7.2 Variation of the interface stress and stretch due to solute segregation

Figure 6 a) demonstrates clearly that the interface stress varies considerably as a function of the solute chemical potential. It has been shown that this can be quantitatively understood in terms of a simple isotropic elastic layer model for grain boundaries.[59] Let the grain boundary be modeled as a layer of material for which the equation of state for $\hat{\mu}(T, x)$ (which determines the 'adsorption isotherm' in the excess formalism) is different from the bulk, so that, in comparison to the bulk, solute is enriched or depleted in the layer. When the concentrations in the layer and in the bulk are independently varied, then the layer will in general tend to be expanded or contracted differently from the bulk. This will give rise to stresses of opposite sign in the two sub-

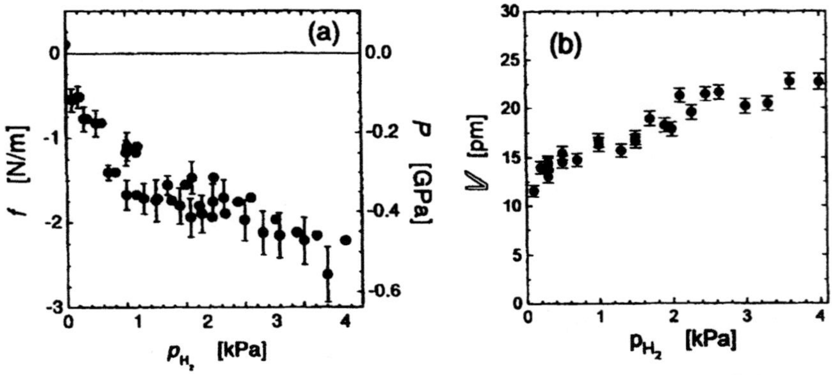

Figure 6. (**a**) Interface stress f (left ordinate) and interface-induced pressure P (right ordinate) in nc-Pd-H versus hydrogen partial pressure p_{H_2}. (**b**) Stretch V versus p_{H_2}. After Ref. 59.

systems, since layer and bulk must remain coherent. The stresses and strains in the layer can be readily analyzed, both for planar[59] and for curved[27,63] interfacial layers. In the limit of small curvature it is predicted that f depends linearly on the specific excess of solute in the layer, according to

$$f = \eta_L t \left(x_V - x_L\right) Y_L / \left(1 - \nu_L\right), \tag{34}$$

where Y and ν refer to Youngs modulus and Poisson number, respectively, and the subscript 'L' denotes layer properties. Among the interesting consequences of the layer model is that the enrichment or depletion of solute in the layer will result in a normal expansion or contraction, respectively, of the layer. This is the equivalent, in the layer model, of the interfacial 'stretch' which emerges independently from the considerations of mechanical equilibrium at interfaces in the theory of continuum mechanics, compare Section 6.1.

The stretch due to hydrogen at grain boundaries can be determined experimentally by measuring separately and in-situ during hydrogen absorption the macroscopic volume change ΔV (with a dilatometer) and the change Δa in the lattice constant (with a diffractometer), a technique which is similar to the famous experiment of Simmons and Baluffi[64] which provided the first direct measurement of the equilibrium vacancy concentration at high temperature. The data for nc-Pd-H [see Fig. 5 a)] can be combined with an equation which is based on the additivity of volume changes,

$$\Delta V = 3\, V\, \Delta a / a + A\, \mathbb{V}, \tag{35}$$

to compute $\mathbb{V}$. Experimental results[60] for the stretch are shown in Fig. 6 b); this first measurement of a deformation of grain boundaries along the normal shows that the internal degrees of freedom for deformation of interfaces can contribute significantly to the macroscopic deformation of nanocrystalline solids.

7.3 Layer thickness and layer properties

Presuming that the layer is isotropic, one finds that the stretch $\mathbb{V}$ and the interface stress f are related by

$$\mathbb{V} = -\left(1 + \nu_L\right) Y_L^{-1} f. \tag{36}$$

Since equation (36) does not depend on the layer thickness, it allows to estimate the value of the biaxial modulus $Y/(1-\nu)$ in the layer based on the data for $\mathbb{V}$ and f. The value is found unchanged relative to bulk Pd.[60]

Interestingly, it has also been found that the partial molar volume of H in grain boundaries in nc-Pd is unchanged relative to H in the bulk.[65] This appears incompatible with the assumptions underlying some theories of grain boundary segregation, in particular the McLean-model. These models derive a mayor contribution to the segregation enthalpy from considerations of the misfit strain energy. The standard assumption is that grain boundaries offer sites with very low partial molar volume for the segregation of interstitials, which implies a considerably reduced misfit strain energy. The data for Pd-H suggest that, after all, misfit strain release may not be so important for grain boundary segregation.

Layer models are also of interest with respect to the issue of potential measures for a 'grain boundary thickness'. Two independent estimates for the thickness of the hydrogen-enriched layer in nc-Pd-H have been brought forward, one based on the jump of the overall solubility[62] and one based on the jump in the interface-induced pressure at the miscibility gap.[27] The first derives 0.9±0.2 nm and the second, 1.1±0.2 nm. An even larger thickness of 2.0 nm was inferred from the variation of the pressure with the solute fraction in the ternary alloy nc-Pd-Au-H.[63] In all cases the width of the segregation layer is considerably larger than the width of the disordered grain boundary core as seen by x-ray diffraction and EXAFS, compare Section 2. This suggests hydrogen enrichment in ordered regions of the crystal lattice near the grain boundary, in agreement with the finding that the biaxial modulus and partial molar volume of H in the layer are unchanged relative to bulk Pd. On the other hand, the width of the enrichment zone is surprising since the microscopic origin for the modified solubility of the grain boundary regions is believed to arise from the modified atomic short-range order, which is localized in the narrow boundary core. To date no convincing explanation for the unexpectedly wide enrichment region has been brought forward.

7.4 Size-dependent miscibility gap

Metal hydrogen-alloys and metal hydrides may serve as model systems for studies of the reversible phase equilibrium in nanoscale materials. It has been reported early on that the miscibility gap of nc Pd-H is narrowed at small grain size.[62,66] A reduced overall solubility is readily explained in terms of layers of material at grain boundaries which do not participate in the phase transformation,[62,67] but a grain-size dependent jump of the lattice constant at the miscibility gap[66] is not readily rationalized in terms of that concept since lattice constant data probe the crystal lattice, not the grain boundary layers. What is more, thin films and multilayers can also exhibit reduced critical temperatures of the miscibility gap, a finding which suggests that the properties of the interior of the crystallites are in some way affected by the reduced scale or by the presence of interfaces. A 'size effect' due to the limitation of critical fluctuations to the finite size of the thin layers has been proposed as a possible explanation,[68] but this is quite unlikely since solid solutions exhibit critical fluctuations only at the critical point of the coherent spinodal, which is typically at least tens of Kelvin below the critical point of the chemical miscibility gap.[69] However, in studies of thin films and multilayers it is generally, at least qualitatively, acknowledged that the chemical equilibrium may be noticeably affected by the stress due to clamping to the substrate.[70] In general, nanocrystalline materials do not interact with a substrate, but we have seen that grain boundaries can induce large stresses in the bulk, and it turns out that these stresses also have a considerable effect on the phase diagram.[27] The situation is similar to that of nanocomposite metal hydrides consisting of crystallites in an amorphous matrix. Such materials exhibit a cooperative dehydriding behavior[71] that is believed to be related to mechanical interactions between the constituent phases, and their phase diagram can be modeled by combining regular solution-type equations of state with an analysis of the stress.[72]

In essence, the modifications in the two-phase equilibrium arise because the interface-induced pressures in the two coexisting phases are different. As a simple example, consider that the grain boundary layer is saturated with solute and acts essentially as an inert layer. By Eq. (34) f will then vary linearly with x_V (since x_L is constant, independent of x_V), and there will be an interface-induced pressure which varies also linearly with x_V. This linear variation was indeed observed experimentally in nc-Pd-Au-H.[63] Assume further that the crystal lattice can be modeled as a regular solution. Then, accounting for the pressure-dependence of μ by Eq. (31),

$$\mu = \mu_1 + 2\,\omega\, x_V + RT\ln\left[x_V/(1 - x_V)\right] + 3\,\eta\,\psi\, x_V/\rho\,, \qquad (37)$$

where the last term arises from the concentration-dependent P, with the concentration-pressure coefficient ψ defined as $\psi = \partial P/\partial x_V|_T$. The important implication of Eq. (37) is that the last term has the same form as the solute-solute interaction term, $2\,\omega\, x_v$; therefore, the interface-induced pressure changes the effective solute-solute interaction energy ω by $\Delta\omega = 3\eta\psi/(2\rho)$. Since the miscibility gap of a regular solution closes at $T_C = -\,\omega/(2R)$ the critical temperature will be modified by the interface-induced pressure. Generalizations of this statement for more realistic equations of state can be derived in the form of a series expansion of μ about the critical point.[27] For nc-Pd-H with a grain size of 9 nm the value $\psi = 4.4$ GPa was measured, and based on this value the reduction in T_C was estimated to be as large as 100 K, see Fig.

7.[27] At any given temperature below T_C, a narrowing of the miscibility gap of the crystal lattice is predicted.

There are many reports of increased solubilities in nanocrystalline alloys and compounds prepared by high energy ball milling. In general these studies do not probe the equilibrium state of the alloy, but it is conceivable that elastic interactions contribute to enhancing the solubility. As yet, there are no direct measurements of the size dependency of T_C in nanocrystalline alloys under conditions of equilibrium.

8. MICROSTRAIN

8.1 Intrinsic nature of the microstrain

The interface-induced stress discussed in Section 6.2 above determines the mean strain and, thereby, the mean lattice constant in the solid microstructure, and we have seen that the associated strain energy contributes intrinsically to the excess free energy of a nanocrystalline material. Besides the *mean* strain, there is also work associated with the local *deviation* of the stress and strain from the mean values. The deviation from the mean can be characterized by the so-called 'microstrain', a term which is mostly used in an operational sense in conjunction with diffraction data, from which it is derived by analysis of the reflection broadening. Before considering the experimental data, we shall shortly demonstrate that microstrain is indeed a necessary consequence of the discrete crystalline structure of nanocrystalline materials.

We shall consider a single grain in a dense nanocrystalline material, and we

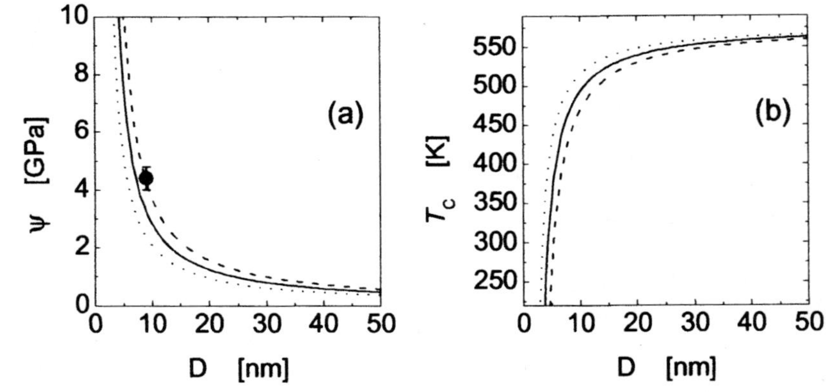

Figure 7. Concentration-pressure coefficient ψ (**a**) and critical temperature T_C of the miscibility gap (**b**) versus grain size D for nanocrystalline Pd-H. Lines: theory using different values for the grain boundary layer thickness t_{GB} (dotted - 0.7 nm, solid - 0.9 nm, dashed - 1.1 nm). ●: experiment. See Ref. 27 for details.

shall ask the question what is the optimum number of atoms in that grain, subject to the constraint that the grain needs to fit - with a minimum amount of strain - into the cavity defined by the neighboring grains (Fig. 8). If our material was amorphous then we could simply add individual *atoms* one by one until the cavity was filled; however, for crystalline matter the task amounts to finding the optimum number of *lattice planes* in any crystallographic direction. Crystallites consist of stacks of lattice planes and their size is an integer multiple of the interplanar spacing. Since the neighboring grains have random orientations the size of the cavity which they define will generally not accommodate an unstrained crystallite. In other words, when adding the last plane one is faced with the dilemma that adding one more plane will generally make the particle slightly too large, so that the lattice needs to be compressed in order to make the particle fit into the cavity, whereas when the last plane is omitted the particle is slightly too small, so that the lattice needs to be strained in tension. Grains in a polycrystalline materials are therefore always strained, and we shall now estimate the magnitude of that strain.

Assume that the boundary plane has the orientation $\langle hkl \rangle$ and that the cavity has the dimension l when measured along the normal to the crystallographic planes; of course, l is of the same magnitude as the grain size, D, and for the purpose of the discussion we may replace l by D. The unstrained grain has the dimension $N\, d_{\mathrm{hkl}}$, where the optimum value for the integer number of planes, N, is the integer part of $(D + d_{\mathrm{hkl}}/2) / d_{\mathrm{hkl}}$. The root mean square (RMS) difference between the dimensions of the strain-free grain, $N d_{\mathrm{hkl}}$, and of the cavity, D, is $d_{\mathrm{hkl}} / (2\sqrt{3})$, and the RMS linear strain along the boundary normal is therefore

$$\langle \varepsilon^2 \rangle^{1/2} = \frac{1}{2\sqrt{3}} \frac{d_{hkl}}{D}. \tag{38}$$

A rigorous discussion of the expectation value for d_{hkl} is beyond the scope of

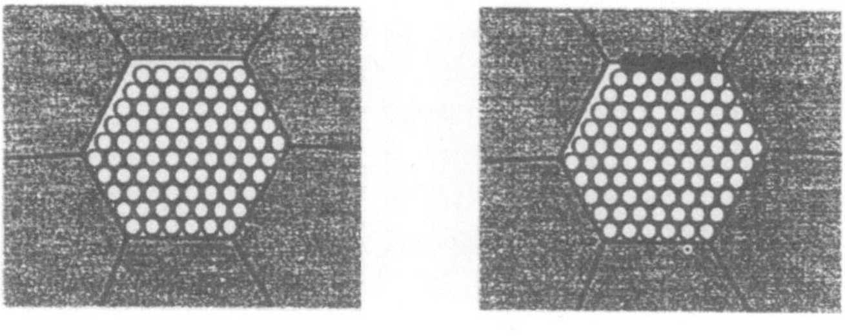

Figure 8. Finding the optimum number of lattice planes in a nanograin. The crystallite on the left is too small, and needs to be strained in tension to be coherent with its neighbors. Adding another row of atoms (on the right) makes the crystallite too large, thereby requiring compressive strain.

this article, but d_{hkl} scales, of course, with the interatomic distance, r_{NN}, and for fcc materials the largest interplanar spacing is $d_{111} = r_{NN}\sqrt{(2/3)}$. High index planes have a long periodicity in the plane; since the period cannot be larger than the grain size the number of possible *hkl* is more and more reduced as the grain size becomes smaller. One can therefore suggest the following statements about the expectation value of d_{hkl}: (*i*) $\langle d_{hkl}\rangle$ is comparable to the near-neighbor interatomic spacing, $\langle d_{hkl}\rangle = c\, r_{NN}\sqrt{(2/3)}$, (*ii*) c is comparable to, but less than unity, and (*iii*) c and, therefore, $\langle d_{hkl}\rangle$ increase with decreasing grain size, since for decreasing D the possible *hkl* are more and more restricted to low index, high d_{hkl} planes. Therefore,

$$\langle \varepsilon^2 \rangle^{1/2} = \frac{c(D)}{3\sqrt{2}} \frac{r_{NN}}{D}. \tag{39}$$

It is instructive to define a characteristic grain size for which the strain reaches a particular value, say 1%, $D_1 = 100\, c\, r_{NN}/(3\sqrt{2})$. Assuming $c = 0.75$ (neglecting the size-dependence of c) and $r_{NN} = 0.275$ nm (the value for Pd), we obtain

$$\langle \varepsilon^2 \rangle^{1/2} = \frac{1}{100} \frac{D_1}{D} \tag{40}$$

with $D_1 = 4.9$ nm. The value for D_1 is in good agreement with experimental data, see below.

8.2 Experimental microstrain

Since the standard methods for determining the grain size of nanocrystalline materials based on integral breadth or Warren-Averbach analysis of wide-angle x-ray diffraction data also supply values for the microstrain there are published data for several nanocrystalline metals and alloys prepared by different methods. Figure 9 a) displays the variation, as a function of D, of the root mean square strain in nc-Pd prepared by inert gas condensation.[36] For samples consolidated at room temperature a least-square fit to the data by the power-law $\langle \varepsilon^2 \rangle^{1/2} = 0.01(D/D_1)^{-n}$ yields n=1.23(6) and D_1=5.4(4) nm. The result for D_1 is in excellent agreement with the predictions in the previous Section, and n>1 agrees qualitatively with the expected size-dependence of $\langle d_{hkl}\rangle$. For a given grain size, the strain was found to be independent of whether the sample was as-prepared or had undergone significant grain growth induced by annealing or aging, supporting the concept of an intrinsic microstrain. This is further supported by the finding that the amount of strain as a function of grain size appears also to be independent of the material and of the method of preparation. This is seen in Fig. 9 b) where the strain data from Ref. 73 for fcc metals prepared by high-energy ball milling (HEBM) are displayed together with the extrapolated fit to the inert-gas condensed Pd from Fig. 9 a). With one exception (Rh), the data for samples prepared by HEBM follow the same relationship between strain and grain size as those for inert-gas condensed Pd.

The empirically universal relation between microstrain and grain size has been argued above to suggest that lattice strain may be an intrinsic property of nanocrystalline solids. Experimentally it is found that, on top of the intrinsic microstrain, additional extrinsic strain can be incorporated in the materials, for instance by deformation.[14,36] Therefore, the intrinsic microstrain represents a lower limit for the amount of lattice distortion.

8.3 Size-dependent interface free energy due to microstrain

It is also of interest to estimate the excess energy associated with the microstrain. Along the direction of deformation the stress due to a uniaxial strain[b] is $S = C\,\varepsilon$, with C a stiffness constant, $C = Y\,(\nu\text{-}1)\,/\,(2\nu^2+\nu\text{-}1)$. The energy density due to the strain is therefore $w = \tfrac{1}{2}\,C\,\varepsilon^2$. If we consider the

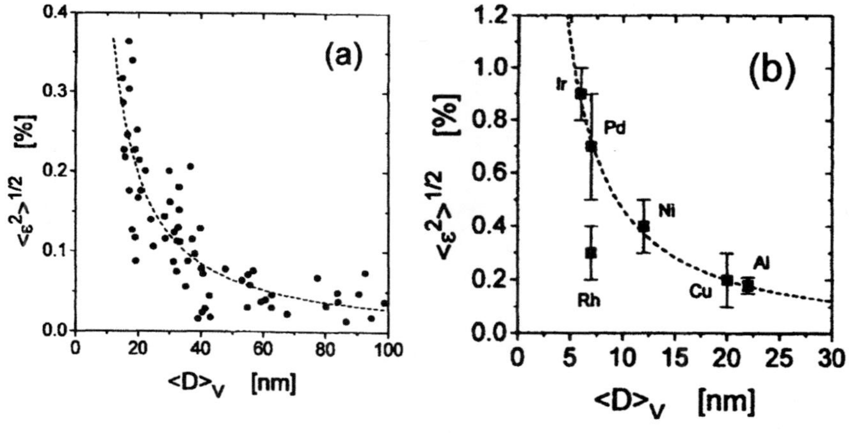

Figure 9. (**a**) Microstrain $\langle\varepsilon^2\rangle^{1/2}$ versus volume-weighted average grain size $\langle D\rangle_V$ for inert-gas condensed nc-Pd (Ref. 36). (**b**) Microstrain (Ref. 73) for other materials, prepared by high-energy ball milling. Dotted lines in (a) and (b): best fit to data in (a) by power law, see text.

---

[b] It is here useful to consider uniaxial strain, not uniaxial stress, because this allows us to treat the strains along the three orthogonal Cartesian coordinates as independent.

strains along the three Cartesian coordinates to be independent and uncorrelated, then the total strain energy is $3w$; with Eq. (40) this evaluates to

$$w = \frac{3\,C}{2 \times 10^4}\left(\frac{D_1}{D}\right)^2 . \tag{41}$$

The contribution, $\Delta\gamma_m$, of the microstrain to the specific Gibbs excess is again obtained as $wV/A = wD/3$, which yields

$$\Delta\gamma_m = \frac{C}{2 \times 10^4}\,\frac{D_1^2}{D} . \tag{42}$$

Continuing to use Pd as the example, with C = 240 GPa, and for $D = D_1 = 4.9\,\mathrm{nm}$ we obtain $\Delta\gamma_m = 0.06$ J/m^2, which is a small fraction of the estimated γ_0 = 1 J/m^2, but much larger than the term γ_f of Section 6.4, and sufficiently large to expect that microstrain energy may contribute measurably to the overall free energy at sizes below 5 nm. More importantly, this term is comparable in magnitude, but opposite in sign, to the reduction of γ due to curvature (Section 4.2), and it may conceivably even overcompensate this reduction, or the equivalent effect of the negative triple line energy.

9. CONCLUDING REMARKS

In conclusion, the thermodynamic functions of materials with a nanometer-scale microstructure differ qualitatively from the ones of their coarse-grained counterparts. There are several, interrelated aspects of this issue: (*i*) When the number of interfaces is large, then the interfacial excess of the extensive variables is no longer negligible compared to their total value; the local properties of the matter at the interfaces will then contribute measurably and, in some instances, decisively, to the macroscopic properties of the material. (*ii*) At small grain size, the total excess per area is no longer a constant, but it varies as a function of the size. Strictly speaking, the total area is then no longer the only appropriate state variable accounting for the effect of interfaces, but curvature, length of linear elements, or the separation between interfaces may need to be considered as additional parameters. (*iii*) At small grain size, long range interactions between interfaces and the bulk may significantly affect the state of the matter not only in the regions near the interfaces, but throughout the entire system. Stress has been discussed as an example, but chemical gradient terms may turn out to be important as well.

There are instances where experimental investigations of reversible changes of state allow the size-dependent thermodynamic properties to be probed directly and quantitatively, or where more qualitative implications are provided by the analysis of irreversible processes in nanomaterials. Such data underline the practical importance of phenomenological analysis and microscopic models of the materials properties at small grain size. In many fields of technology, the lateral size of the elements of the microstructure keeps decreasing, and an understanding of the fundamental thermodynamics, in combination with a data base for the interfacial materials parameters, will be increasingly relevant for future optimization of their processing and performance, as well as the understanding of failure mechanisms of such nanoscale materials and devices.

Future studies may exploit the potential for controlling the overall properties of nanocrystalline materials through the manipulation of the nanoscale microstructure and of the interfaces even further by designing nanomaterials with tunable properties. This has been suggested based on the growing body of evidence implying that space-charge layers at interfaces can have a significant influence on a variety of physical properties (see, for instance, the contribution by J. Maier in this volume). Space-charge layers may arise as a consequence of the different electronegativity of the abutting phases in heterophase interfaces, but of particular interest is the notion that a suitable experimental set-up may allow to induce deviations from charge neutrality by an externally applied voltage. This may open the way to future nanocrystalline materials in which the structure and electronic properties can be tuned by varying the applied voltage.[74]

ACKNOWLEDGMENT

Support by the Deutsche Forschungsgemeinschaft (SFB 277, Heisenberg-Program, and Center for Functional Nanostructures) is gratefully acknowledged.

REFERENCES

1. Gleiter, H., in *'Deformation of Polycrystals: Mechanisms and Microstructures'*, Proc. 2nd Risø International Symposium on Metallurgy and Materials Science, edt. N. Hansen, A. Horsewell, T. Leffers, and H. Liltholt (Roskilde, DK 1981), 15.
2. Wolf, D. and Yip, S., editors, *'Materials Interfaces - Atomic Level Structure and Properties'* (Chapman & Hall, London 1992).

3. Sutton, A.P., and Balluffi, R.W., *'Interfaces in Crystalline Materials'* (Clarendon Press, Oxford 1995).
4. Gibbs, J.W., *'On the Equilibrium of Heterogeneous Substances'*, Trans. Conn. Acad. III (1878); also in: *'The Collected Works of J.W. Gibbs'*, Vol. I (Longmans, Green, and Co. , New York 1928), Section 'Theory of Capillarity', pp. 219 ff.
5. Fowler, R.H., and Guggenheim, E.A., *'Statistical Thermodynamics'* (Cambridge University Press, Cambridge, UK 1960).
6. Cahn, J.W., in *'Interfacial Segregation'*, edt. W.C. Johnson and J.M. Blakely (ASM, Metals Park, OH 1979), 3.
7. Greer, A.L., in *'Mechanical Properties and Deformation Behavior of Materials having Ultra-Fine Microstructures'*, edts. M. Nastasi, D.M. Parkin, and H. Gleiter (Kluwer, Dordrecht 1993), 53.
8. Würschum, R., Schaefer, H.E., Chapter 11 in *'Nanomaterials: Synthesis, Properties, and Applications'*, edts. A.S. Edelstein and R.C. Cammarata (Institute of Physics, Bristol, UK 1996).
9. Weissmüller, J., Chapter 10 in *'Nanomaterials: Synthesis, Properties, and Applications'*, edts. A.S. Edelstein and R.C. Cammarata (Institute of Physics, Bristol, UK, 1996).
10. Weissmüller, J., Proc. 22nd Risø International Symposium on Materials Science, (Roskilde, DK 2001), in press. http://www.risoe.dk/afm/symp22.
11. Birringer, R., Gleiter, H., Klein, H.P., Marquardt, P., Phys. Lett. 102A (1984), 365.
12. Gleiter, H., Marquardt, P., Z. Metallkunde 75 (1984), 263.
13. Tschöpe, A., Birringer, R., J. Appl. Phys. 71 (1992), 5391.
14. Valiev, R.Z., Islamgaliev, R.K., Alexandrov, I.V., Prog. Mater. Sci. 45 (2000), 103.
15. Löffler, J., Weissmüller, J., Phys. Rev. B 52 (1995), 7076.
16. Fitzsimmons, M., Eastman, J., Müller-Stach, M., Wallner, G., Phys. Rev. B 44 (1991) 2452.
17. Schlorke, N., Weissmüller, J., Dickenscheid, W., Gleiter, H., NanoStruct. Mater. 6 (1995), 593.
18. Boscherini, F., de Panfilis, S., Weissmüller, J.; Phys. Rev. B 75 (1998), 3365.
19. Phillpot, S.R., Wolf, D., Gleiter, H., J. Appl. Phys. 78 (1995), 847.
20. Keblinski, P., Wolf, D., Phillpot, S.R., Gleiter, H., Scripta mater. 41 (1999), 631.
21. Keblinski, P., Phillpot, S.R., Wolf, D., Gleiter, H., Acta mater. 45 (1997), 987.
22. Van Swygenhoven, H., Farkas, D., Caro, A., Phys. Rev. B 62 (2000), 831.
23. Weissmüller, J., NanoStruct. Mater. 3 (1993), 261.
24. Hillert, M., in *'Lectures on the Theory of Phase Transformations'*, edited by H.I. Aaronson (AIME, New York 1975).
25. Weissmüller, J., J. Mat. Res. 9 (1994), 4.
26. Ehrhardt, H., *'Thermische Stabilität nanostrukturierter Materialien'*; Dissertation, Univ. Saarbrücken 1998. Shaker Verlag, Aachen 1998.
27. Weissmüller, J., Lemier, C., Phil. Mag. Lett. 80 (2000), 411.
28. Cahn, J.W., J. Chem. Phys. 66 (1977), 3667.
29. Cahn, J.W., Acta Metall. 9 (1961), 795.
30. Cahn, J.W., and Hilliard, J.E., J. Chem. Phys. 28 (1958), 258.
31. Gurtin, M.E., Murdoch, A.I., Arch. Rat. Mech. Anal. 57 (1975), 353.
32. Thomson, W. (Lord Kelvin), Phil. Mag. 25 (1887), 503.
33. Weaire, D., Phelan, R., Phil. Mag, Lett. 69 (1994), 107.
34. Levine, D., Grest, G.S., Phil. Mag, Lett. 74 (1996), 303.
35. Günther, B., Kumpmann, A., Kunze, H.-D., Scripta Met. 27 (1992), 833.
36. Weissmüller, J., Löffler, J., Kleber, M., NanoStruct. Mater. 6 (1995), 105.
37. Lu, K., Sun, N.X., Phil. Mag. Lett. 75 (1997), 389.
38. Srinivasan, S.G., Cahn, J.W., Jónsson, H., Kalonji, G.; Acta mater. 47 (1999), 2821.

39. Caro, A., van Swygenhoven, H., Phys. Rev. B 63 (2001), 134101.
40. Palumbo, G., Thorpe, S.J., Aust, K.T., Scripta Metall. Mater. 24 (1990) 1347.
41. Morgan, F., and Taylor, J.E., Scripta metall mater. 25 (1991), 1907. Taylor, J.E, Interface Science 7 (1999), 243.
42. Herring, C., in *'Structure and Properties of Solid Surfaces'*, edts. R. Gomer and C.S. Smith (Univ. of Chicago Press, Chicago 1953), 5.
43. Gärtner, F., Bormann, R., Birringer, R., Tschöpe, A., Scripta Mater. 35 (1996), 805.
44. Villain, S., Cabané, J., Knauth, P., Scripta Mater. 38 (1998), 1003.
45. Krill, C.E., Klein, R., Janes, S., Birringer, R., Mat. Sci. Forum 179-181 (1995), 443.
46. Weissmüller, J., Krauss, W., Haubold, T., Birringer, R., Gleiter, H., NanoStruct. Mater. 1 (1992), 439.
47. Terwilliger, C.D., Chiang, Y.-M., Acta Metall. 43 (1995), 319.
48. Malow, T.R., Koch, C., Mater. Sci. Forum 225-227 (1996), 595.
49. Ruckenstein, E., Chi, J.C., J Chem. Soc. Faraday Trans. II 71 (1975), 1690.
50. Tadros, T.F. in 'Surfactants in Solution' vol. 3, edt. K. Mittal (Plenum Press, NY 1984).
51. Weissmüller, J., Ehrhardt, H., Phys. Rev. Lett. 81 (1998), 1114.
52. Ibach, H.; Surf. Sci. Reports 29 (1997), 193 .
53. Cammarata, R.C., Prog. Surf. Sci. 46 (1994), 1.
54. Spaepen, F., Acta mater. 48 (2000), 31.
55. Shuttleworth, R., Proc. Phys. Soc. A63, (1950), 444.
56. Gurtin, M.E., Weissmüller, J., Larché, F., Phil. Mag. A 78 (1998), 1093.
57. Weissmüller, J., Cahn, J.W., Acta mater. 45 (1997), 1899.
58. Hoffmann, M., *'Grenzflächenspannung in nanokristallinem Pd'*; Dissertation, Univ. Saarbrücken 1998. Birringer, R., and Hoffmann, M., in preparation.
59. Weissmüller, J., Lemier, C., Phys. Rev. Lett. 82 (1999), 213.
60. A different version of Eq. (30) was proposed by Q. Jiang and coworkers (Q. Jiang, L.H. Liang, and Z.S. Zhao, J. Phys. Chem. B 105 (2001), 6276.). Their derivation neglects the work of deforming the bulk, and the result is therefore representative of the interfacial free energy density φ, not the excess free energy γ.
61. Larché, F., Cahn, J.W., Acta Metall. 21 (1973), 1051.
62. Mütschele, T., Kirchheim, R., Scripta Met. 21 (1987), 135.
63. Weissmüller, J., Lang, C., Lemier, C., Scripta mater. 44 (2001), 1899.
64. Simmons, R.O., Balluffi, R.W., Phys. Rev 117 (1960), 62.
65. Mütschele, T. Dissertation, Universität Göttingen 1989.
66. Eastman, J.A., Thompson, L.J., Kestel, B.J., Phys. Rev. B 48 (1993), 84.
67. Pundt, A., Sachs, C., Winter,M., Reetz, M.T., Fritsch, D., Kirchheim, R., J. Alloys Comp. 293-295 (199), 480.
68. Song, G. , Geitz, M., Abromeit, A., Zabel, H., Phys. Rev. B 54 (1996), 14093.
69. Cahn, J.W., Acta metall. 9 (1961), 795.
70. Weidinger, A., Nagengast, D., Rehm, CH., Klose, F., Pietzak, B., Thin Solid Films 275 (1994), 48.
71. Orimo, S., Fuji, H., Ikeda, K., Acta mater. 45 (1997), 331.
72. Tessier, P., Schulz, R., Ström-Olsen, J.O., J. Mater. Res. 13 (1998), 1538.
73. Eckert, J., Holzer, J., Krill, C., Johnson, W., J. Mater. Res. 7 (1992), 1751.
74. H. Gleiter, J. Weissmüller, O. Wollersheim, R. Würschum, Acta mater. 49 (2001), 737.

DIFFUSION IN NANOMATERIALS

Jean Bernardini and Dezsö L. Beke*

Laboratoire Matériaux et Microélectronique de Provence associé au CNRS, UMR 6637, Faculté des Sciences de St. Jérôme, 13397 Marseille cedex 20, France.
**Department of Solid State Physics, University of Debrecen, H-4010 Debrecen, P.O. Box 2. Hungary.*

1. INTRODUCTION

Grain boundaries are, generally, diffusion short circuits; consequently, the major part of material transport will occur by grain-boundary diffusion in nanomaterials where a large amount of atoms can lie on grain or interphase boundaries (50% for a grain size equal to 5 nm; 20% for a grain size equal to 10 nm). The development of nanolayered materials and bulk nano-materials prepared e.g. by compaction of nanocrystalline powders is now raising the question of how the overall mass transport changes as the microstructural scale is reduced to nanometer dimensions.

It is well known from classical treatments of grain-boundary or interface diffusion that there are three different grain-boundary diffusion regimes: type A, B and C. In type C regime the diffusion takes place only along the grain- boundaries, in B regime there is also diffusion into the bulk and in type A regime the bulk diffusion fields are overlapping. We will see that for grain sizes, d, in the order of 10 nm the B regime practically can not be observed and either A or C regimes will be realized. In type A regime a significant enhancement of intermixing or solid state reactions is observed, with a high effective interdiffusion coefficient (e.g. in surface alloying). On the other hand in many cases the process will take part dominantly along grain- or phase boundaries (type C regime) leading to phenomena such as degradation of multilayers by grain-boundary grooving, pinhole formation and coarsening, or solid state phase transformations in thin films. Furthermore, in hetero-diffusion experiments by radio tracers (i.e. when the tracer is different from the matrix atoms) segregation coefficients describing the matching conditions between the diffusion source and the grain-boundary (as well as between the grain-boundary and the free surface, if there is a terminal free surface present in the experiment) should be also taken into account.

Another interesting question raised during the interpretation of the already existing data on grain-boundary diffusion in nanocrystalline materials: Whether the grain-boundary diffusion coefficients measured in these alloys are identical to those obtained in the microcrystalline state or not? An answer to this question would solve a fundamental point concerning the structure of GB in these materials: is it a well defined and ordered one as in coarse grained materials or a frozen-gas like structure. We will see that there is an increasing number of new experimental evidences that the above diffusion coefficients agree very well with each other, i.e. in most of the cases the structure of relaxed grain-boundaries in nanocrystalline and polycrystalline samples is very similar. At first sight one could say that the diffusion coefficient value and thus the nature of GB seems to depend on the technique applied for the synthesis of the material more than on the grain size. In fact, the main point is to compare nano/coarse grain samples where GBs are near an equilibrated state from the point of view of the structure and the chemical composition.

Diffusion in nanostructures presents challenging features even if the role of structural defects (dislocations, phase- or grain-boundaries) can be neglected. This can be the case for diffusion in amorphous materials or in epitaxially grown, highly ideal thin films or multilayers where diffusion along short circuits can be ignored and "only" principal difficulties, related to nanoscale effects, raise. For example for diffusion in such crystalline materials, one of the most important differences – as compared to diffusion along long distances (orders of magnitude longer than the atomic spacing, a) – is that the continuum approach can not be automatically applied. Another important feature is the gradient energy correction to the driving force for diffusion. This correction becomes important again if large changes in the concentration take place along distances comparable with a, and results in an additional term in the atomic flux, proportional to the third derivative of the concentration. In interdiffusion of phase separating systems this can lead not only to intermixing but to spinodal-like modulations as well.

A further difficulty can be related to the role of the diffusion induced stresses. In many treatments of diffusion (at least in metallic systems), it is supposed that the relaxation of stresses, proportional to stress free strains created by the differences in atomic volumes and intrinsic diffusion coefficients, is fast and complete (see e.g. Darken's treatment of interdiffusion). However, for short diffusion distances, the time of relaxation of stresses can be longer than the time of diffusion. Thus fading back effects of stresses can become important and cannot be ignored anymore.

Sometimes – e.g. in diffusion in MBE grown thin films or multilayers – the diffusion distances can be shorter than the typical distance between the sources and sinks of point defects (diffusion vehicles). While in "normal" diffusion the equilibrium concentrations of such defects are set by the sources or sinks (and the characteristic distance, l, between them is much shorter than the macroscopic diffusion length, L), if $L<l$ the mechanism of diffusion can change, similarly as it was already observed for diffusion in dislocation-free Si single crystals.

At short diffusion distances in interdiffusion, leading to formation of reaction products, the usual parabolic law of diffusion can be violated, if the role of reactions at the interface is taken into account. At short diffusion times, when the gradient of concentration is large, the reaction controlled rate of the mass transfer can be slower than the diffusion current and a *linear* growth law can be observed.

It is also well-known, that the nucleation of a new phase (e.g. from a supersaturated solution or from an under-cooled amorphous phase) always needs a certain critical size, d_c, which is typically in the nanometer range. If, for example, in a solid-state diffusion reaction the width of the diffusion zone, L, is less than d_c the given phase cannot form. In case of more than one diffusion product, this can lead to different interesting observations in the first stages of the heat-treatments.

The above examples exposed some of the most challenging features of diffusion in nanostructures. In this review, we will try to cover the present status of this rapidly growing field.

2. DIFFUSION IN PRESENCE OF GRAIN BOUNDARIES

In presence of structural defects (dislocations, grain boundaries, phase boundaries....), mass transport is generally increased. Most mathematical treatments of GB diffusion start from Fisher's model developed for a single GB [1].

2.1. Fisher's model

As it can be seen in Fig.2.1., according to this model, the GB with a thickness δ is perpendicular to the surface where the diffusion source is deposited; it is considered as an isotropic static medium, semi-infinite, uniform, characterised by a higher diffusivity as compared to the bulk. With these assumptions and considering the case when there is also diffusion into the grains (type B regime), the Fisher model is based on the following statements:

1. Volume and grain-boundary diffusion coefficients (D_v and D_b) are isotropic, concentration, position and time independent.

2. The concentration and the flux of the diffusant are continuous at the boundary/volume interface.
3. The grain-boundary thickness is so small that the concentration variation of the diffusant in the boundary is negligible in the y direction.

Then

$$\left| \begin{aligned} \frac{\partial c_v}{\partial t} &= D_v \left(\frac{\partial^2 c_v}{\partial x^2} + \frac{\partial^2 c_v}{\partial y^2} \right) \text{ for } |y| \geq \frac{\delta}{2} \\ \frac{\partial c_b}{\partial t} &= D_j \frac{\partial^2 c_b}{\partial x^2} + \frac{2D_v}{\delta} \left. \frac{\partial c_v}{\partial y} \right|_{|y|=\frac{\delta}{2}}, \end{aligned} \right| \tag{2.1}$$

where c_v and c_b are the concentrations of the diffusing species in the bulk and in the boundary, respectively, and t the time.

One can find several solutions of these equations in the literature. For example, Suzuoka [2] gave the solution for the case of a thin-film source. Whipple's solution [3] applies when the surface concentration is maintained constant on the surface of the sample. Le Claire [4] formulated a simple

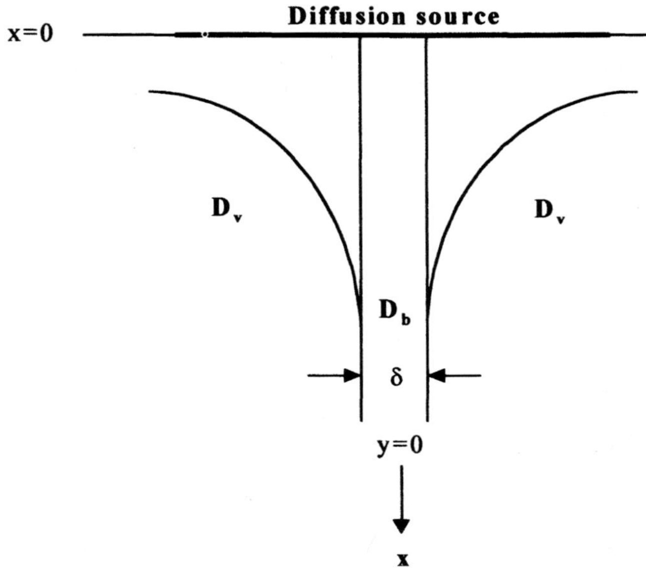

Figure 2.1. Illustration of Fisher's model

relationship between δD_b, D_v, t and the slope of the experimental penetration curve ($\partial \ln \bar{c} / \partial x^{6/5}$, where $\bar{c}$ is the value of the concentration in the yz plane). These expression are used, in general, for the interpretation of experimental penetration curves. Plotting the logarithm of the concentration versus $x^{6/5}$, at penetrations larger than $5\sqrt{D_v t}$, a straight line is obtained, the slope of which can be used for evaluation of δD_b.

However, the solutions of equations concern only diffusion in bicrystals. Levine and MacCallum [5] treated the grain-boundary self-diffusion for polycrystals in the most general way. They considered that the crystal contains randomly oriented grain boundaries and that the grain size distribution is also random and could show that Le Claire's equations can be used in this case as well. Therefore, these equations give the basis of the mathematical analysis of grain-boundary diffusion in polycrystals.

Up to now, most (GB) self-diffusion experiments have been carried out in coarse grained metallic polycrystals using the Fisher model [6,7]. However, over the last years, some GB diffusion measurements have been also carried out in different nano- or sub-micro-crystals, where the effect of GB migration can be also important.

2.2. Kinetics regime of diffusion in static and mobile GBs

Three atomic fluxes may be present in a polycrystal. The first one, from the free surface into the volume, and the second one, out of GBs into the adjacent grains, are described by the volume diffusion coefficient D_v; the last one, inside the grain boundaries, is described by the GB diffusion coefficient D_b.

2.2.1. Diffusion in polycrystals containing stationary GBs

Depending on i) the relative magnitude of D_v and D_b ($D_b \gg D_v$), ii) the volume diffusion length $L=(D_v\ t)^{1/2}$, iii) the grain diameter d, and iv) the GB width δ, Harrison [8] has proposed to distinguish three kinetic regimes (called A, B, and C) in polycrystals assuming parallel and stationary GBs.

- In the C kinetic regime ($\alpha = k\delta/2(D_v\ t)^{1/2} \gg 1$, where $k = C_b/C_v = \exp(-\Delta H_s/RT)$ is the equilibrium GB segregation factor and ΔH_s the segregation energy), the volume diffusion is negligible and atomic transport occurs within the boundaries. Analytically, the system can be described in the same way as for volume diffusion by replacing D_v with D_b.

- In the B kinetic regime ($\alpha \ll 1$ and $(D_v\ t)^{1/2} \ll d$), GB diffusion takes place with simultaneous volume diffusion from the boundary into the adjacent crystal and $(D_v\ t)^{1/2}$ is much lower than the spacing between GBs; the total concentration of tracer c in a section parallel to the diffusion direction y is the sum of two contributions c_1 and c_2: c_1 is due to direct volume diffusion from the surface while c_2 results from the leakage of the tracer from GB into the volume. The diffusion profiles consist of two parts; from the second part, which varies linearly with $y^{6/5}$ for penetration depths larger than $5(D_v\ t)^{1/2}$, one can deduce the triple product ($k\delta D_b$) using different solutions of Fick's equations [1].

- In the A kinetic regime $(D_v\ t)^{1/2} >> d$, the different diffusion zones overlap each other, resulting in a macroscopic homogeneous diffusing distribution which appears to obey Fick's law as for a homogeneous system. The effective diffusion coefficient (D_{eff}) is equal to an average of D_v and D_b weighted by the ratio of the number of diffusing atoms in the grains to that in GB [9]

$$D_{eff} = g\ D_b + (1-g)\ D_v, \qquad (2.2)$$

where g is the grain-boundary volume fraction ($g \approx \delta/d$; the factor of proportionality depends on the grain shape, but is in the order of unity). If $d \cong 10$ nm, then only the first term will be dominant, since usually $D_b/D_v > 10^4$.

Now, we can rewrite the condition for the B kinetic regime as

$$100\ \delta k < (Dt)^{1/2} < d/2 \qquad (2.3)$$

from which it can be seen that if *d≅10 nm this condition can not be fulfilled.* Thus according to the conditions for the type A and B regimes ($(Dt)^{1/2} \gg d$ for A and $20(Dt)^{1/2} < \delta$ for C) – depending on t – *either A or C regimes will be observed.* Thus, according to [10], in any technically important interdiffusion or solid state reaction process in nanomaterials, a high effective diffusion coefficient will characterize the rate (e.g. in surface alloying) in type A diffusion. On the other hand, in many cases the process will take place dominantly along grain- or phase boundaries (type C regime) leading to phenomena such as degradation of multilayers by grain-boundary grooving, pinhole formation and coarsening [11,12], or solid state phase transformations in thin films [13].

It should be pointed out that in Harrison' s paper the regimes are determined in terms of strong inequalities. In fact, the limit for the occurrence of the regime A seems to be lower than that given by Harrison ($(D_v\ t)^{1/2} > d/2$) [14], while the limits on α for the occurrence of type C and B kinetics are less restrictive ($\alpha > 10$ and $\alpha < 0.01$ respectively). So it remains a large domain of α values ($0.01 < \alpha < 10$) where all the classical solutions of Fick's equations relative to C and B regimes cannot be used. Recently, Mishin and Razumoskii [15, 16] as well as Mishin and Herzig [17] have proposed a more sophisticated classification of the diffusion regimes, with a mathematical analysis, for isolated GB as a function of the values of the parameters α and β (*Table 1*).

Table 1.
GB diffusion regimes in a stationary bicrystal, after Mishin and Razumovskii [15,16]

Regime	C	B_1	B_2	B_3	B_4
Value α	>> 1	≈ 1	<< 1	<< 1	<< 1
Value β	>> 1	>> 1	>> 1	≈ 1	<< 1
Measured parameter	D_b	$k\delta D_b$ and $k\delta$	$k\delta D_b$	$k\delta D_b$	D_v

The B2 regime corresponds to the B regime in Harrison's classification. The B4 regime considers the case when at high temperature diffusion from the surface into the volume can predominate over GB diffusion ($\beta = kD_j\delta / 2D_v(((D_v\ t)^{1/2} \ll 1$); the B1 and B3 are transition regimes between C/B2 and B2/B4. The most interesting regime which appears in this new classification is the B1 transition regime ($\alpha \sim 1$), which can exist when one tries to reach type C conditions (see also [18,19]).

The above analysis concerns only bicrystals and coarse grained polycrystals when $(D_v\ t)^{1/2} < (D_b\ t)^{1/2} < \delta$. Depending on the interrelations between the four parameters $k\delta$, $L=(D_v t)^{1/2}$, $L_b=(D_b t)^{1/2}$ and δ, eight kinetic regimes can be in fact realized in polycrystals [17]. The C' and B'$_2$ regimes introduced in [17] are analogue to the classical C and B2 regimes respectively, but i) the GB penetration depth L_b is much larger than δ in the C' and B'$_2$ regimes and ii) a numerical factor q, which depends on the grain shape, is present in the parameter $P' = qk\delta D_b$ determined in the B'$_2$ regime. In the A' regime a factor q' is also present in the effective diffusion coefficient $D_{eff} = q'k\delta D_b /d$ deduced from the penetration profile. At last, in the A_o regime, which implies very strong GB segregation of the diffusant and /or very small grain size d, the GB diffusion coefficient D_b can be measured.

2.2.2. Diffusion along a single moving GB

In presence of a grain boundary moving with a velocity v, other diffusion regimes must be introduced according to the magnitude of the variables α and $\gamma = vt/(D_v\ t)^{1/2}$ [20-22]. The definitions of these different kinetic regimes are given in Table 2 without considering the transition regimes B1 and B3.

Table 2
Diffusion regimes in a bicrystal with stationary/moving GB, after Mishin and Razumovskii [16]

Regime	C	B_2	MC	M B_2	B_4
Value α	>> 1	<< 1	>> 1	<< 1	<< 1
Value β	<< 1	<< 1	>> 1	>> 1	
	$vt \ll (D_v t)^{1/2}$ $(D_v t)^{1/2} \ll \delta$	$vt \ll (D_v t)^{1/2}$ $\delta \ll (D_v t)^{1/2}$	$(D_v t)^{1/2} \ll vt$ $vt \ll \delta$	$(D_v t)^{1/2} \ll vt$ $\delta \ll vt$	

A complete mathematical analysis of the diffusion along a single GB, perpendicular to the surface of the specimen and moving with a constant velocity v, was done by Mishin and Razumovskii [21]. Looking at the different solutions of Fick's equations [1, 21], one can see that neglecting the effects of GB migration in measurements by the classical sectioning technique can lead to large errors in the diffusivity values. As an example:

- In presence of type MC kinetic regime and significant boundary displacement ($vt > 10\ \delta$), the plot $\ln C = f(y)$ is linear [21, 23] with a slope equal to $(v/\delta\ D_b)$. This behavior, very different from the case of type C kinetics, can be misinterpreted as type B2 diffusion along stationary boundaries.

- Under the conditions of type MB2 kinetic regime (which practically exists for $\gamma > 6$), the plot $\ln C = f(y)$ is linear, with a slope equal to $(v/\delta\ D_b)^{1/2}$. Thus i) the product (δD_b) can be only obtained if v is known, ii) using the classical Fisher solution relative to type B2 regime [24], one will determine an apparent value of (δD_b), underestimated by a factor of $(\pi\gamma/2)$.

2.2.3. Diffusion in polycrystals with stationary and moving GBs

A mathematical analysis of GB diffusion in polycrystals taking into account the migration of some GBs has been published [25]. It assumes that i) a fraction f of GBs moves with the same time independent velocity (v), ii) the other GBs remain stationary, iii) all the GBs remain perpendicular to the surface, and iv) the diffusion takes place in the MB2 kinetic regime (vt and $(D_v t)^{1/2} \ll d$). The diffusion can be described using the models relating to a single stationary and a single moving GB. The tracer profile consists of three parts (Fig.2.2) with the presence of an intermediate step between the near surface part, associated with volume diffusion, and the deeper penetrating zone associated to diffusion along stationary boundaries. The whole profile (except for $(D_v t)^{1/2}$ near the surface) can be described by the equation:

$$\text{Ln}\ C = \text{Ln}\ [q_1 \exp(-q_2\, y^{6/5}) + q_3 \exp(-q_4\, y)] \tag{2.4}$$

where the meaning of q_i is graphically indicated in figure 2.2. It is true that some assumptions seem to be simple; however, the model i) clearly indicates that diffusion measurements in presence of moving GB require high-sensitive detectors allowing to extend the tracer detection as far from the surface as possible, and ii) is very useful to understand the shape of experimental profiles observed in refractory [25,26] or very high purity metals.

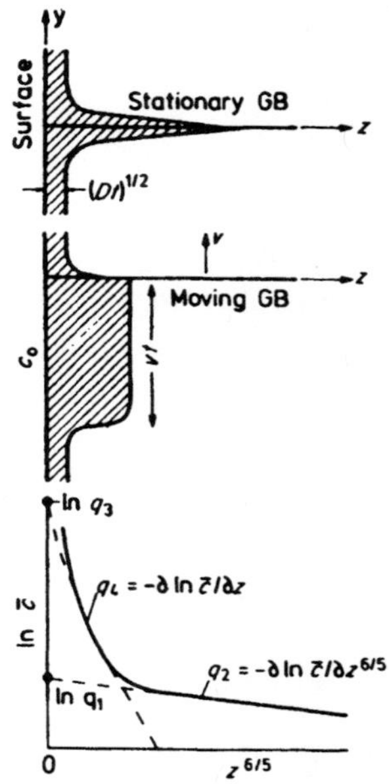

Figure 2.2. Schematic distribution of diffusing atoms for a stationary (A) and moving (B) GB and expected shape of the GB depth penetration profile (C) in presence of stationary and moving GBs, after Guthoff et al. [25].

2.3. GB segregation and shape of the penetration profiles

The effect of segregation on GB diffusion in polycrystals has been already discussed in details [1, 27-29]. We will just recall here the effects of non-linearities, i.e. the effect of i) GB saturation and ii) interactions of diffusing species on the penetration plots as they are surely exacerbated in nanomaterials.

On the basis of Fisher's model, Gibbs has demonstrated that the GB width δ is replaced by the product $k\delta$ in experiments other than self diffusion ones without any change in the shape of the diffusion profiles [30]. However, one must assume that the so called Henry isotherm is valid, i.e. $C_b = kC_v$, which means that i) GB saturation (Mc Lean's isotherm) and ii) interactions between species (Fowler-Guggenheim's isotherm), which can be respectively written in the forms:

$$C_b = k\,C_v / (1 - C_v + kC_v) \tag{2.5}$$

and

$$C_b = k\,C_v \exp[\exp[\gamma (2\,C_b - 1)](2\,C_b - 1)] / (1 - C_v + k\,C_v \exp[\gamma (2\,C_b - 1)]) \tag{2.6}$$

(where $\gamma = \Delta H_{mix}/RT$ and ΔH_{mix} is the enthalpy of mixing in GB), are not taken into account.

The consideration of the saturation and interaction effects imply to solve Fick's equations by numerical methods. This has been done with the simplifying conditions as used by Fisher, using either Mc Lean's isotherm [31-33] or Fowler-Guggenheim's isotherm [32]. In both cases the predicted profiles exhibit a non linear behaviour in Fisher's coordinates $Ln(C_b)$ and y. They present near the surface a curvature which depends on segregation energy and enthalpy of mixing. Then they become linear at large penetration depths with the same slope as in the case of Henry's isotherm. It is assumed by the authors that the type and the extend of the curvatures will be similar on the ln C versus $y^{6/5}$ plots resulting from the more precise solutions of Fick's equations given by Whipple [3] or Suzuoka [4]. These predictions have been observed for silver diffusion in copper [34] where the curvature of the profiles due to the non-linear silver segregation extends up to 1500 $(D_v t)^{1/2}$, while it disappears for the diffusion of silver in copper-silver alloys (where GBs are pre-saturated before the diffusion of the radiotracer).

More recently, Bokstein *et al.*[35] attempt to take into account a more realistic description of GBs than the slab model (i.e. an energetic distribution of GB sites leading to a variation of the segregation energy from one GB site to another) not in a discrete atomistic, but in a continuous thermodynamic manner. They expressed the relationship between C_b and C_v in the form:

$$C_b = \int [k\, C_v / (1 - k\, C_v - C_v)]\, w(k)\, dk \tag{2.7}$$

The terms within the brackets correspond to the Mc Lean isotherm at the GB sites of type i (caracterized by a segregation energy ΔH_i^s) and w(k) dk refers to the distribution of the sites on the heats of interaction with the tracer. The curvature of the profiles depends on the maximum (ΔH_S^{max}) and minimum (ΔH_S^{min}) values chosen for ΔH_s, but the GB inhomogeneity always amplifies segregation (and thus decreases the rate of diffusion) when compared with the saturation effect applied to an homogeneous GB (Fisher's model). However, it exists in polycrystals a distribution of GB's leading to a distribution of D_b [36]. Thus their experimental separation is a difficult task.

One can expect however that all the segregation effects (and particularly the effect of GB saturation) will be less important in nanostructured materials than in coarsened polycrystals. In effect, owing to the number of GBs and the synthesis conducted under UHV conditions, the grain boundaries must be purer in the first type of materials if one excepts the contamination during the preparation (specially by oxygen) and supersaturated metastable nanocrystalline solid solutions. Thus decreasing the grain size there will be a change of the character of the GB segregation isotherm from Fowler type to Mac Lean type without any saturation (see as an example Fig. 1b in [37]).

2.4. Effects of stresses on GB diffusion

In a set of more recent papers of Bokstein and his co-workers [38,39,40], it was illustrated that the mechanical stress field (both the stress gradient and its hydrostatic component) can have a considerable influence on the grain boundary diffusion in thin films. Even the effect of the stress gradient and the effect of the hydrostatic component can be separated as "gradient effect" and "mobility effect" (see also later). In [38], the authors argue that the mechanical stress can reach as high values as 1 GPa and the small film thickness can lead to large gradients as well. Furthermore, at low temperatures the stress relaxation can be hindered and the hydrostatic component of the stress field can also enhance or suppress the grain boundary diffusion (across the exponential dependence of the mobility on the pressure). They gave detailed solutions of this problem for B- and C-kinetic regimes of grain boundary diffusion and established the theoretical basis of future, desired experiments. At the moment, there is a lack of unequivocal evidences for the existing order of magnitude differences of the grain boundary diffusion coefficients measured, e.g. in thin film made on different substrates (and thus being in different stress states), although the above explanation offers a plausible solution.

2.5. Experimental methods allowing the determination of GB diffusion data

Direct methods allowing to determine GB diffusion data have been extensively analysed by Kaur, Mishin and Gust [1]. To summarise they can be divided mainly in two categories: the first ones involve the measurements of the diffusant concentration profile inside the specimen after the diffusion treatment (sectioning methods); the second ones are based on the continuous detection of the diffusant on the back surface, after penetration throughout finite-thickness samples (surface accumulation methods). Many experimental techniques are available for carrying out these measurements, depending on the experimental conditions and the thickness of the samples. Radiotracers associated with different microsectioning techniques must be used for self diffusion experiments in case of sectioning methods. For impurity diffusion their main advantage is to allow diffusion coefficients determination under conditions of infinite dilution without changing the chemical composition. SIMS (or AES sputter-etching technique) can be also used to determine bulk and GB diffusion coefficients in these conditions at lower temperatures [41], but it is very often difficult to avoid the effect of the sputter damage on the profiles. In contrast, surface accumulation methods using surface-analytical techniques, such as Auger electron spectroscopy, are very well adapted to study GB diffusion at low temperatures in thin specimen in type C kinetics regime. Hwang *and* Balluffi. [42] analysed the problem in an array of uniformly spaced parallel grain boundaries in a thin-film system. They assumed that the atoms arriving from the grain boundaries spread out on the back surface. The geometry and the notations of the thin-film system are illustrated in figure 2.3.

The Hwang-Balluffi relation, which links the grain-boundary diffusion coefficient D_b with the average concentration c_s of the accumulation surface, is given by:

$$(k''/k')c_s/c_0 = 1 - \exp(-\omega t') \qquad (2\text{-}8)$$
$$\omega = \delta D_b \lambda / \delta_s h k'.$$

where λ is the grain-boundary density (*e.g.* for a polycrystal having cubic grains with d, $\lambda = 2/d$), k' is the segregation coefficient at the accumulation surface/grain-boundary interface and k'' the segregation coefficient at the grain-boundary/source interface. They are defined by the proportions c_s/c_b and c_b/c_0, respectively. It should be pointed out that the form given here is not exactly that written in [42]; its a little bit generalised version [43]. The t' quantity is a "corrected time" of the form:

$$t' = t - t_0 \qquad (2\text{-}9)$$

where t_0 is a constant, taking into account that a transient phenomenon occurs in each diffusion measurement before a quasi-steady-state is reached and t is the real time. The Hwang-Balluffi equation can be applied under the following conditions: i) a quasi-steady-state grain-boundary diffusion current to the accumulation surface has been established; ii) the surface diffusion rate is sufficiently rapid so that the segregated atoms are uniformly distributed laterally in the surface region; iii) a constant concentration of diffusing source atoms is maintained in the grain-boundary/source interface.

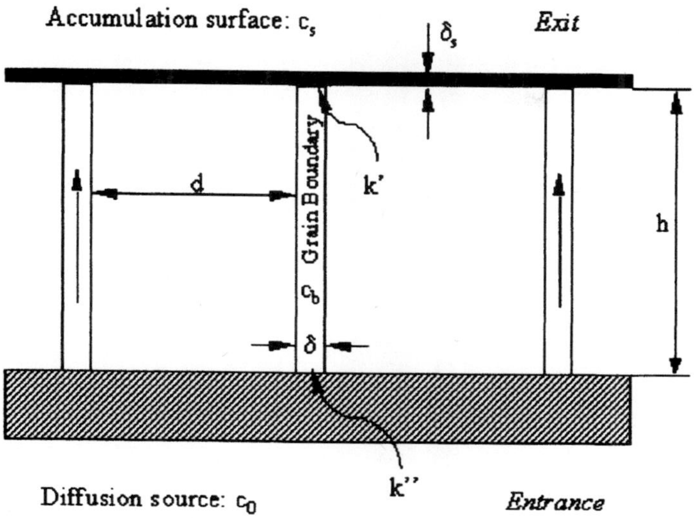

Figure 2.3. Geometry of the thin film system used in the Hwang-Balluffi method

2.6. GB diffusion measurements in nanocrystals and structure of GBs

Few experiments have been performed in nano or sub-micro crystals where the time scale of the diffusional heat treatment must be short enough compared to times of relaxation to expect time independent diffusivity owing to the metastability of the materials.

It should be pointed out that an accurate measurement of the diffusion coefficients implies samples without internal stress, open porosity and cracks plus a correct knowledge of the diffusion regime to take into account a possible overlapping of diffusion fluxes from adjacent GBs and a possible migration of GBs. Moreover let us recall that, in classical polycrystals, reproducible results and agreement between works performed in different laboratories in the last twenty years are mainly due to a good preparation of the samples before the diffusion heat treatment (by annealing the samples first at high temperature and then at the intended diffusion temperatures for a time long enough to expect structural and chemical equilibration of GBs). So, all the methods of synthesis of nanomaterials are not suitable for diffusion measurements and much of the work has suffered from the difficulty of making specimens that are fully dense and large enough to make traditional diffusion measurements. Let us just recall that among the different ways allowing to obtain nanostructures, i) inert gas condensation followed by in situ compaction at room temperature, and mechanical allowing/ milling lead generally to the presence of nanovoids (which in case of vacuum compaction may be as large as the grain size); ii) the crystallisation-prepared samples, free of pores, are generally dense and pure, but can show amorphous intergranular layers; iii) nanostructured metals prepared by severe plastic deformation enable to fabricate samples free of pores with a grain size slightly higher than 100 nm, but with a very high density of defects and a high level of stresses which will relax at the beginning of the diffusion treatment. In contrast thin films can be produced without any porosity, with different grain sizes (as a function of the thickness), on different type of supports; they appear to be well adapted to compare diffusion in nano- and polycrystals and to study the effect of stresses if any. So we will not present here results obtained in the early 90's in samples produced by the inert gas condensation of vapours and in situ powder compaction, where diffusivities as high as the diffusivities on free surfaces (for reviews see [44], [45]) appears to be linked to a high density of internal free surfaces [46-49]. One must bear in mind in effect that it is difficult to retain a nanocrystalline structure and obtain full densification. Hot extrusion and hot isostatic pressing requires use of high pressures and elevated temperatures during a long time. In ceramics in particular, 1-9GPa of pressure is required to prepare good nanocrystalline samples and such a pressure cannot be used owing to the classical thickness of the specimen. In the same way, we will not detail the high diffusion coefficients measured in samples prepared by severe plastic deformation (for a review see ref. [50] and [51]). Obviously the results will depend on the value of the relaxation time of the different defects with respect to the diffusion time in these samples, which are far from the equilibrium state. As an example, recent results reported for copper diffusion in nickel [52], coarse grained (CG), and nanostructured samples prepared by severe plastic deformation (SPD) or electrodeposition (ED) show that i) the coefficient of diffusion measured in SPD nickel at 423K decreases a lot when the sample is pre-annealed for one hour at 523K and cannot be measured (as in CG samples) when the pre-treatment is performed at 623K(cf. table 3).

Table 3

Copper GB diffusivities in different types of (99.95%) nickel samples at 423K, after Kolobov *et al.* [51]

samples	Grain size (μm)	Pre annealing	Diffusion time	D ($m^2.s^{-1}$)
CG	20	1h at 1073K	extrapolation	4.3 10^{-19}
SPD	0.3	1h at 398K	3h	9.6 10^{-15}
SPD	0.3	1h at 523K	50h	2.8 10^{-18}
SPD	0.3	1h at 623K	50h	Not measurable
ED	0.03		5h	3.8 10^{-17}

The main interest of this work is to show on the first hand that the grain size, as indicated in the introduction, is not the main parameter, as diffusion in ED samples (not pre-annealed) is lower than that in SPD samples pre-annealed at 398K. Different methods of synthesis may produce different types of GBs. There is for example some evidences that mechanical attrition creates GBs with a higher energy than that present in equilibrated high-angle GBs [53]. On the other hand, Bolotov's results clearly demonstrate the necessity to anneal the samples to obtain quasi-equilibrated GBs before to perform diffusion measurements. A change in the GBs atomic density in Ni_3Fe nanocrystalline samples has been observed from Mössbauer and small angle neutron scattering measurements after heat treatments performed at 100°C [54]. A systematic study of the diffusion as a function of the pre-treatments in the same type of samples prepared by different methods and a comparison with the behaviour of the same material produced by classical rolling up to 90% (but without any recrystallization annealing at high temperatures) would be a way to determine the role of GB relaxation on the diffusion and to obtain results in relaxed GBs.

Table 4

Diffusion studies in nanocrystals

SYSTEM	PREPARATION	Size (nm)	d (%)	TECHNIQUE	T(K)	REMARKS	Ref.
Cr/Al	Thin film	30	100	SIMS and AES	732 ; 823		[59]
Ni_2Si	200 nm Ni deposited on Si	40-60	100	Formation of Ni_2Si thin films	523-573	Effect of orientation	[60]
Al-Mg3%	Equal-Channel Angular Pressing (ECAP)	700		Electron-probe microanalysis	523-773	No evidence for the occurrence of enhanced diffusion after ECAP.	[61]
Fe/Fe	Evaporation-condensation Compaction 3GPa 400K in UHV + explosive compaction	18 20	87 95	Radiotracer Sputter-sectioned by Ar^+	450-500	D_b values are one order of magnitude higher than in conventional samples and decrease when d is increased.	[62]
Pd/Fe	Cluster compaction Severe plastic deformation	20 100	98	RBS Radiotracer	423-523	The high density of the samples and preannealing treatments lead to diffusivities similar to those in conventional samples.	[63]
Cu/Ag	Bilayers prepared by magnetron sputtering	20	100	AES Hwang-Balluffi method	393-428	Fair agreement with radiotracer measurements in conventional samples.	[43]

In contrast with the early diffusion studies, where very high diffusion coefficients were measured, as early as 1992-1995:

i) Quantitative X-ray diffraction studies of nanocrystalline Pd indicate that there is no significant grain boundary excess volume in this material [55].

ii) Calorimetric analysis of the grain growth in nanocrystalline copper prepared by compacting small particles lead to grain boundary energy values similar to those found in conventional polycrystalline copper [56].

iii) Average widths of GBs similar in nano and classical polycrystals were measured by the Mössbauer technique in several different systems prepared by mechanical attrition [57,58] (approximately 0.5 nm and 1 nm in FCC and BCC alloys respectively).

All these experimental evidences are in agreement with lower diffusivities, slightly higher than that in classical GB, reported more recently in the literature, using more dense materials obtained by electrodeposition, innovative methods of consolidating powders or dense thin films. These studies are summarised in table 4. They indicate that the diffusivity, and thus the GBs structure, can be similar in nano-size materials and in conventional polycrystals, when the density is not far from the theoretical one and when pre-annealing treatments can relax the grain boundary structure.

3. DIFFUSION ON NANOSCALE

As it was mentioned in the Introduction, principal difficulties could also arise in materials free of structural defects (amorphous materials or epitaxially grown, highly ideal thin films or multilayers), just because the diffusion length can become comparable to the atomic spacing, a. In these cases the continuum description of diffusion (i.e. the classical forms of the Fick I. and II. laws) can not be automatically applied. The problems related to the above question can be classified into two groups:

i) diffusion on a discrete lattice
ii) diffusion in amorphous materials.

3.1. Diffusion on a discrete lattice

3.1.1. Effect of the strong concentration dependence of the diffusion coefficients

It was shown already decades ago (see e.g. [64, 65]) that, even if the diffusion coefficient, D, is independent of concentration, for diffusion distances, L, shorter than 10a, the discrete model has to be applied. In order to illustrate this, let us consider the discrete model of diffusion proposed by Martin [66] for an exchange mechanism in a binary AB alloy. This consists of a slab of material with N lattice planes, normal to the y axis. Each atom in a plane (e.g. in the *i*th atomic layer) has z_l nearest neighbours in this plane as well as z_v in plane $(i+1)$ and plane $(i-1)$. Thus the coordination number is $Z= z_l +2z_v$. The change in relative concentration of A or B in plane i is given by

$$dc^K_i/dt = J^K_{i-1,i} - J^K_{i,i+1} \qquad K=A,B, \qquad (3.1)$$

where $J^K_{i,i+1}$ is the net flux per lattice site of K atoms from layer *i* to *i+1*. Let us now define $J^B_{i,i+1} = J_{i,i+1}$ as:

$$J_{i,i+1} = z_v\{c_i(1-c_{i+1})\Gamma_{i,i+1} - c_{i+1}(1-c_i)\,\Gamma_{i+1,i}\}. \tag{3.2}$$

Here e.g. c_i is the fraction of B atoms in the *i*th layer, and $\Gamma_{i,i+1}$ gives the exchange frequency of a B atom in layer i with an A atom in layer i+1. Accordingly, for concentration independent jump frequencies (and if driving forces other than the gradient of concentration are zero) $\Gamma_{i,i+1}=\Gamma_{i+1,i}=\Gamma$, the Fick II. law has the form:

$$\partial c_i/\partial t = \Gamma(c_{i+1}+c_{i-1}-2c_i). \tag{3.3}$$

The term in the bracket, divided by d^2 (where d is the distance between the atomic planes), can be rewritten to the second derivative of the relative concentration. Indeed it can be obtained - if, as it is illustrated in Fig 3.1., the surface concentration, $n= cd/\Omega$ (Ω is the atomic volume), is a smoothly varying function of the position - by Taylor expansion and stopping at the second derivatives. Now, it is easy to show [64, 65] - again for the case of a concentration independent diffusion coefficient $D=\Gamma d^2$ - that taking a sinusoidal initial concentration profile ($c\sim\sin\{2\pi x/\lambda\}$), the solutions of the continuum Fick equation and eq. (3.3) will be identical only if $\lambda>10d$.

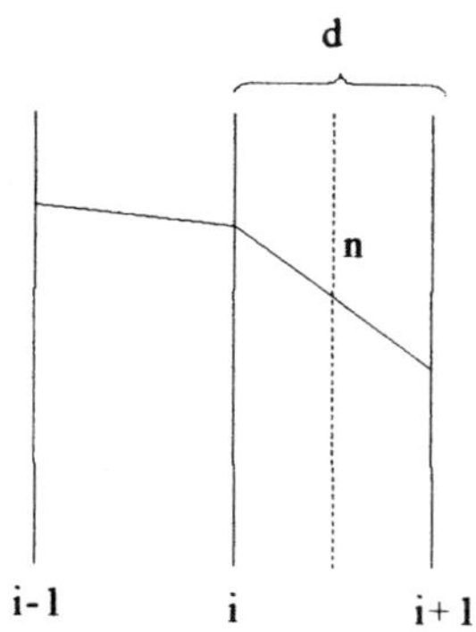

$$nd = c$$

$$n_i = n - \frac{d}{2}\frac{\partial n}{\partial x} + \frac{(d/2)^2}{2!}\frac{\partial^2 n}{\partial x^2} - \frac{(d/2)^3}{3!}\frac{\partial^3 n}{\partial x^3} + K$$

$$\frac{\partial c}{\partial x} = \frac{c_{i+1} - c_i}{d}$$

$$\frac{\partial^2 c}{\partial x^2} = \frac{\dfrac{c_{i+1} - c_i}{d} - \dfrac{c_i - c_{i-1}}{d}}{d}$$

Figure 3.1. Change of the (continuous) surface and volume concentration across atomic planes and its derivatives.

However, in the case of a strongly concentration dependent interdiffusion coefficient (i.e. in the presence of a strong non-linearity) the validity limit of the continuum approach can be shifted to larger values by about one order of magnitude [67], i.e. it will be in the order of 10 nm. Fig. 3.2 shows the dependence of the distance (the critical modulation length in a multilayer), above which the continuum and discrete model give the same results, as a function of the parameter, m', determining the exponential concentration dependence of the diffusion coefficient:

$$\log D = \log D(o) + m'c. \qquad (3.4)$$

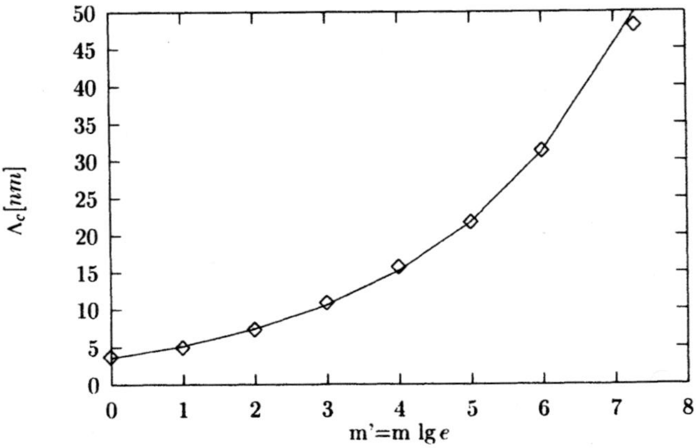

Figure 3.2. Values for the critical modulation length, Λ_c, above which the continuum model is valid, as a function of the value of exponent in the concentration dependence of D.

These results were obtained in [67] by carrying out calculations for diffusion intermixing in multilayers of ideal solid solutions, both from the Martin model and from the finite difference calculations of the continuum Fick II. equation with diffusion coefficients equivalent to that used in the atomic model. Indeed, for this concentration dependent diffusion coefficient in the Martin model an appropriate choice of the activation energies $E_{i,i+1}$ and $E_{i+1,i}$ of $\Gamma_{i,i+1}$ and $\Gamma_{i+1,i}$ can be such [66, 67, 68, 69] that

$$E_{i,i\,i+1} = E^o - \alpha_i + \varepsilon_i \qquad \text{and} \qquad E_{i+1,i} = E^o - \alpha_i - \varepsilon_i \qquad (3.5)$$

where E^o is the saddle point energy and α_i and ε_i are proportional to V_A-V_B and $V = V_{AB} - (V_A + V_B)/2$, respectively ($V_A$, V_B and V_{AB} are the pair interaction energies and V has its usual meaning). Obviously, α_i and ε_i are also functions of $c_{i,-1}$, c_i, c_{i+1} and c_{i+2}:

$$\alpha_i = [z_v(c_{i-1} + c_{i+1} + c_i + c_{i+2}) + z_l(c_i + c_{i+1})](V_A - V_B)/2 \qquad (3.6)$$

and

$$\varepsilon_i = [z_v(c_{i-1} + c_{i+1} - c_i - c_{i+2}) + z_l(c_i - c_{i+1})]V. \qquad (3.7)$$

For ideal solid solutions the interaction energy, V, is zero and thus $\varepsilon_i = 0$. Furthermore, applying the same Taylor expansion as before, but assuming that the second derivatives are negligible (the concentration is a smoothly varying function of position), it would be easy to show that $\alpha_i \sim c(V_A - V_B)$, which corresponds to (3.4) (see also below).

There was another interesting feature of the results obtained in [67], which is illustrated in Fig. 3.3. For strong concentration dependence of D (i.e. for strong non-linearity), the interface between the two materials remains atomic sharp and shifts as a whole until the component with small D has not been consumed. The calculations were carried out for the Mo-V system (with m'=7.3) and it can also be seen that the diffusion is very asymmetrical: there is a fast dissolution and diffusion of Mo into V, but there is no diffusion in Mo. This behaviour, as it is illustrated in Fig. 3.4, was indeed observed very recently in amorphous Si/Ge system by Auger-depth profiling technique [70].

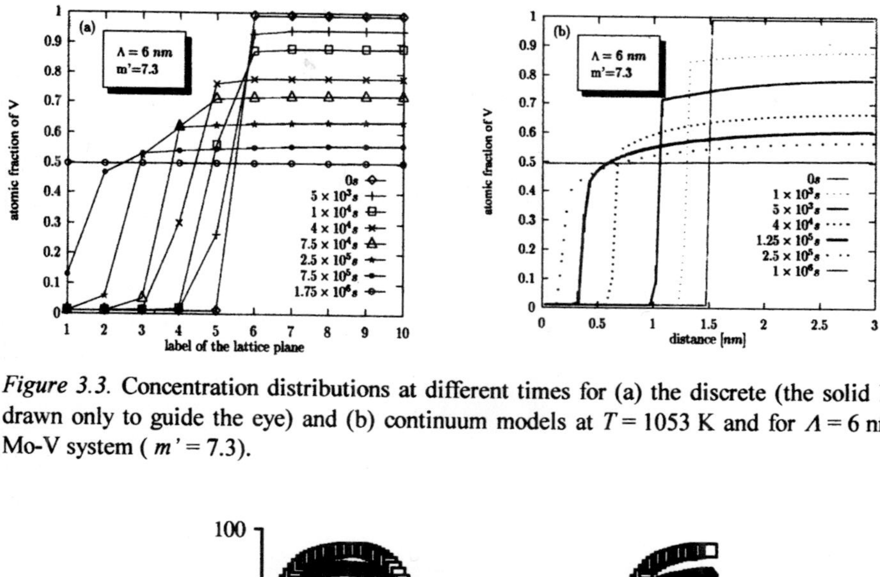

Figure 3.3. Concentration distributions at different times for (a) the discrete (the solid lines drawn only to guide the eye) and (b) continuum models at $T = 1053$ K and for $\Lambda = 6$ nm in Mo-V system ($m' = 7.3$).

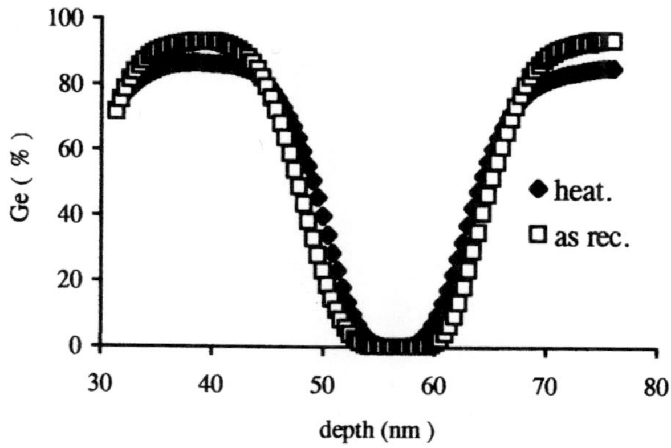

Figure 3.4. One period of the measured Auger depth profiles for the as-received and annealed (at 680 K for 100 h) amorphous Si/Ge multilayer [70]. It can be seen that there is an increase of the Si content in Ge, but no Ge appears in Si as well as the Si layer shrinks.

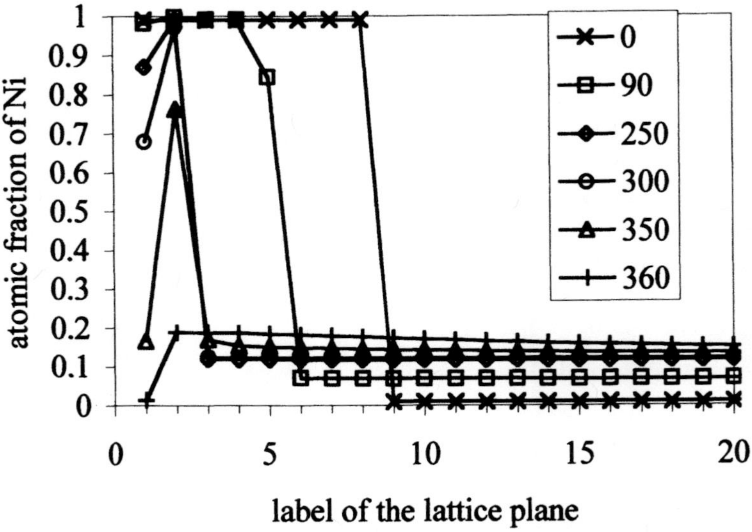

Figure 3.5. Time evolution of concentration profiles in Ni/Cu(111) discrete lattices: for 8 monolayers of nickel and 51 layers of Cu (of which only 12 are shown here). Time units, shown in the insert, are given in special units used in the calculations [71].

This feature can lead to even more interesting results if we have dissolution of the thin film into a substrate [71]. Fig. 3.5 shows the results of the simulations carried out for Ni dissolution into Cu (again the system is ideal, i.e. V=0). It can be seen that the dissolution starts at the interfacial layer, and until this is not consumed, the next layer remains complete (on the curve related to 90 times units only the concentration of the 5th plane changes while the 4th and 6th planes remain almost pure Ni and Cu, respectively). Thus, the interface shifts step by step. This layer-by-layer dissolution takes place until the moving "interface" reaches the Ni layer just before the last. Then, due to the driving force for surface segregation, the intermixing will be continued by the saturation of Cu in the top layer. The change in the second layer will be retarded according to the segregation isotherm. Finally, after the saturation of the surface layer by Cu, the final homogenization takes place by the complete dissolution of the second Ni layer.

Turning back to the dissolution, it can be seen that the dissolution of the (i-1)-th plane begins only after the complete dissolution of the i-th plane. This means – if the substrate is semi-infinite and the diffusion coefficient depends strongly on the concentration [71] – that this layer-by-layer dissolution results in a periodic behaviour as a function of time: each plane dissolves subsequently reproducing the same process. Therefore, the rate of the interface shift (and thus D) is also a periodic function of time and the time evolution within one period characterizes the dissolution of one individual plane. But then it is worthwhile to define an average value over T (T is the time necessary for the consumption of one layer) characterizing the mean speed of the interface shift. Due to the step-by-step character of the dissolution, this average speed of the interface shift, v, should be constant, independent of time. This result is inherently related to the strong non-linearity of the problem: the strong concentration dependence of the diffusion coefficient (or $\Gamma_{i,i+1}$) shifts the validity limit of the continuum approach (from which a parabolic law i.e. $v \propto t^{-1/2}$ would be expected) out of the range considered here. Of course, after the dissolution of more and more layers the supposition that $c_{i+1} \cong 0$ will be less and less valid and we will have a transition to the parabolic dissolution. Obviously, this transition will depend on the concentration dependence of the diffusivity. Fig. 3.6. shows the position of the interface versus time, obtained from simulation for a semi-infinite Cu(111) substrate with 100 atomic layers of Ni. Due to the oscillating character, mentioned above, the curve has periodic oscillations around the straight line fitted. On the other hand, the slope of the straight line is $1 \pm 8\times10^{-4}$, i.e. the average shift is indeed linear.

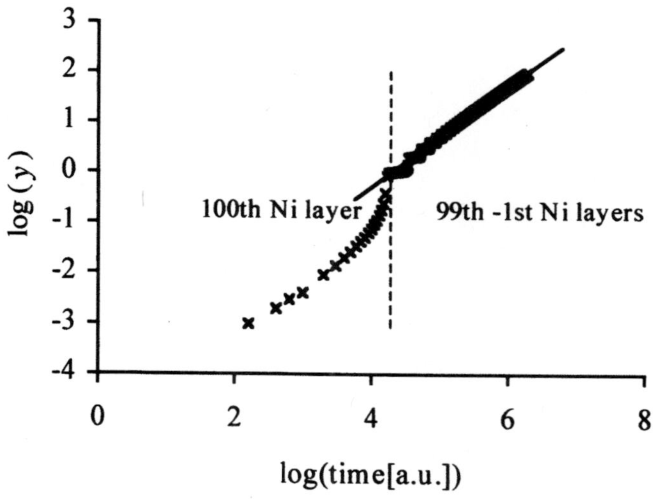

Figure 3.6. Position of the interface versus time for the dissolution of 100 Ni layer into the semi-infinite Cu(111) substrate (see also the text).

Although the oscillating character of the dissolution – because of technical difficulties - could not be resolved experimentally in [71], the above nice simulation result was also confirmed by measuring the kinetics of the Auger signals of Ni and Cu from the top of 8 monolayers Ni. Fig. 3.7. shows the final results for the average time evolution of the Ni thickness (given in monolayer units) versus time at 679 K. It can be seen that *n* is a linear function of time in the stage of the layer-by-layer dissolution. Note that usually the first couple of points fall above the straight line fitted to the first part of the *n(t)* functions.

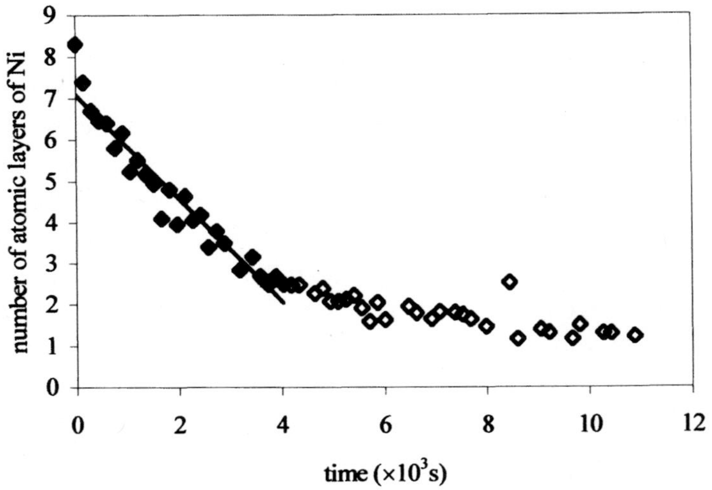

Figure 3.7. Time evolution of the Ni thickness at 679 K. The first part of the curve fitted by a straight line corresponds to the part controlled by interdiffusion.

Since the samples were prepared at room temperature the reason of this misfit can be re-crystallization and relaxation of the surface layers during the heat treatment. Furthermore, the change in the slope at the second layer is a clear evidence of the effect of surface segregation.

3.1.2. Gradient energy corrections

It can be seen from (3.6) and (3.7) that α_i can be rewritten – using $Z = z_l + 2z_v$ as well as the relations $(c_{i-1}-c_i)-(c_{i+1}-c_{i+2}) = (d^2)\partial^2 c/\partial x^2$ and $c_i + c_{i+1} = (d/2)\,\partial^2 c/\partial x^2 + 2c$ (see also Fig. 3.1.) – as [69]:

$$\alpha_i = (V_A - V_B)\{cZ + (z_v + Z/4)d^2\partial^2 c/\partial x^2/2\} = cZ(V_A - V_B) + \alpha_i'. \tag{3.8}$$

Thus

$$\Gamma_{i,i+1} = \Gamma_i^h \exp[(\alpha_i' - \varepsilon_i)/kT]$$

and (3.9)

$$\Gamma_{i+1,i} = \Gamma_i^h \exp[(\alpha_i' + \varepsilon_i)/kT],$$

with Γ_i^h, being the jump frequency in a *homogeneous* alloy,

$$\Gamma_i^h = \nu \exp[-(E^o + cZ\{V_A + V_B\})/kT], \tag{3.10}$$

and in an ideal system ($V=0$, $\varepsilon_i=0$): $\Gamma_{i,i+1}=\Gamma_{i+1,i}=\Gamma_i=\Gamma_i^h\exp[\alpha_i'/kT]$.

Thus (3.2) can be rewritten as

$$J_{i,i+1}= z_v\Gamma_i\{c_i(1-c_{i+1})\exp(-\varepsilon_i/kT) - c_{i+1}(1-c_i)\exp(\varepsilon_i/kT)\}. \quad (3.11)$$

If $\varepsilon_i/kT \ll 1$ then $\exp(\varepsilon_i/kT)\cong 1+\varepsilon_i/kT$, and now keeping terms in the Taylor expansion of the concentration up to the *third* order (i.e. for example $c_i - c_{i+1} = -d\partial c/\partial x - (d^3/24)\partial^3 c/\partial x^3$), after some algebra (see also [69,71,72]), one has:

$$\mathbf{J}_{i,i+1}= J_{i,i+1}d^2/\Omega= -D_i(\partial c/\partial x)/\Omega + D_i[2\kappa/f_o'' - d^2/24](\partial^3 c/\partial x^3)/\Omega. \quad (3.12)$$

Here, according to the usual definition $D_i=z_v d^2\Gamma_i/\Omega$; κ and f_o'' are the gradient energy coefficient and the second derivative of the free energy of the homogeneous system [73].

This expression – except the second term proportional to d^2 in the bracket – is similar the well known Cahn-Hilliard expression [72,74,75]. However, as it was already illustrated by Martin [72], the classical Cahn-Hilliard expression contains a diffusion coefficient belonging to a *homogeneous* alloy. On the other hand, as it can be seen from (3.8) - again keeping terms up to the third order – in the exponent of D_i, α_i' is proportional to the second derivative of the concentration, i.e. inhomogeneity corrections lead not only to the term proportional to the third derivative of concentrations, but also to an additional factor in D_i !

Furthermore, the role of the term proportional to d^2 should also need some comment. Even for an ideal solution, where $V=0$ and thus κ is also zero, the presence of this term, because of its negative sign, would lead to decomposition for concentration modulations with a wavelength shorter than the order of d. This is obviously an artefact of the simplifications made during the transition from the discrete model to the continuous one and not surprising at all (see the comments on the validity limits of the continuum approach as well). In fact, this correction term corresponds to that derived in [76] for diffusional homogenisation of multilayers, and is given in the form as a correction to κ.

Nevertheless the role of these refinements would need more attention if the diffusion takes place on a nanoscale and one describes it by continuum equations.

3.2. Diffusion in amorphous materials

In the light of the above results, it is a challenging and open question how to describe the diffusion in an amorphous material on nanoscale. One would guess that deviations from the classical Fick laws are also expected in this regime, especially if sharp concentration gradients are present and if the diffusion coefficients strongly depend on the concentration. One of the most plausible questions would be whether any linear shift of the sharp interface would be observed or not, i.e. the result observed in Ni/Cu system is the consequence of the non-linearity or it is related to the discrete character of the crystalline lattice as well? Until now there are no such experimental investigations in the literature, although our results on amorphous Si/Ge system indicate (see again Fig. 3.4) that the situation is at least similar in that sense that the originally sharp interface remains sharp and shifts keeping its shape [70].

Another interesting question is that this shift of the interface is "only" the effect of the deterministic character of the discrete equations (Martin's model) used in our calculations or not. Our preliminary Monte Carlo simulations for a *discrete* lattice indicated that the linear shift of the interface can be the case at the beginning of the process (i.e. the smoothening of the interface, due to the fluctuations, is not significant) if the concentration dependence of the diffusion coefficient is strong (i.e. if it changes at least by four orders of magnitude between the two sides of the diffusion zone) [77].

3.3. Stress effects

Some of the effects of the stresses on grain-boundary diffusion have been already treated in Section 2. Here again we want to consider effects of stresses of diffusional origin on the bulk diffusion. The first difficulty is that almost all of the sophisticated theories (except some trials [78,79]) are developed in the framework of a continuum model [80,81,82], and thus conclusions e.g. on the interplay of the stresses and the non-linearity can be treated in this approximation only.

Nevertheless, we can have some guidelines for qualitative considerations. We suppose that at the beginning there are no stresses present, and the diffusion currents are different (in the previous considerations for the non-linearity the effects appear also if the intrinsic diffusion coefficients were equal to each other). In interdiffusion, in general, because of the different partial volumes and mobilities there is an imbalance of the volume transport. This is equivalent to the creation of a non-uniform stress-free strain [80-84] (and thus stress): on one side of the diffusion couple contractions, while on the other side extensions will arise. In the classical Darken analysis [85], it was assumed that the relaxation of stresses is fast and complete and only a convective transport (Kirkendall-shift), as a result of this relaxation, was included into the theory.

In reality, the internal stress developed during the process contributes to the thermodynamic potential for diffusion of atomic species, leads to plastic deformation and gives rise to an elastic contribution as well. The convective transport – as a result of the plastic flow, realised either by creep or dislocation glide or by both of them – obviously relaxes the internal stress and this will lead to a feeding-back effect in the coupled set of equations of the system [80].

The presence of a stress (or strain) field in general can be described by a tensor, σ; its presence can influence the atomic currents across the chemical potential, μ. Since $\mu(\sigma)$ can have different forms for different sink/source densities of point defects determining the diffusion, and for different boundary conditions, *the final form of the currents - as a consequence of the long range character of elastic fields - can also be dependent on the boundary conditions [84].*

For example, if there are no sinks or sources present in the sample than $\mathrm{div}\mathbf{J_j}=0$, and since the atomic currents has the form

$$\mathbf{J_j} = -L_{jj}\mathrm{grad}\mu_j(\sigma), \tag{3.13}$$

the equation

$$\Delta\mu_j(\sigma) = 0. \tag{3.14}$$

should be solved. For a uniformly, non-hydrostatically stressed solid [86]

$$\mu_j(\sigma) = \mu_{jo} - V_j\sigma; \qquad V_j\sigma = V_{kl}\sigma_{kl}\mathbf{n}_k\mathbf{n}_l, \tag{3.15}$$

where $\mathbf{n}_k$ is the unit surface normal vector (for a vacancy, V has a negative sign). Thus

$$\mathrm{grad}\mu = - V_j\mathrm{grad}\sigma, \tag{3.16}$$

i.e. $\mathbf{J_j}$ depends on the geometry of the sample (boundary conditions). This type of treatment can lead to the classical description of creep [87]: vacancies can be annihilated and created only at the free surfaces (or grain boundaries), which leads to a continuous change of shape. The description of interstitial diffusion, or solute- redistribution (Gorsky effect) as treated by Cahn [88,89] as well as by Cahn and Larche [90,91] or by Baranowsky [92] can be also made by this way. In these cases usually the stress field can be expressed as the function of the concentration, i.e. the $\sigma(c)$ function can be given and *no relaxation of stresses are allowed.* Usually a second order non-linear equation describes the time evolution of the system. For example, following Cahn's treatment [88,89] it was shown for homogenisation in multilayers by Spaepen [93] that the decay curve of the intensity of the first order small angle X-ray satellites, I, can be characterised by an effective interdiffusion coefficient

$$D_{eff}=D_D\{1 + [2\delta^2E/(1-\nu)f_o'']\,[1-(I_o/I)^{1/2}]\}. \tag{3.17}$$

Here $\delta=(da/dc)/a$ (a is the lattice parameter), E is the Young modulus, ν is the Poisson ratio and D_D is Darken's interdiffusion coefficient. It can be seen that D_{eff} is time dependent: it decreases and vanishes when the increase of the strain energy stops the diffusion at a terminal value of I, given by the condition $D_{eff}=0$.

In another limiting case, it can be supposed that *the relaxation of stresses are allowed* and there is a high density and efficiency of sources/sinks (of vacancies) [94]. In vacancy quasi-equilibrium, for a cubic system:

$$\mu_j(\sigma) = \mu_{jo} - V_j\sigma_{jj}/3 = \mu_{jo} + V_jp. \tag{3.18}$$

On the basis of this equation the current due to the presence of a pressure gradient can be easily handled (across (3.13.) [80,81]), *although again different boundary conditions can lead to different long-range stress fields and thus to different currents*. Furthermore, different stress fields can lead to different elastic and plastic deformations, and also to different hydrostatic pressures, which - across the Arrhenius-type pressure dependence of the mobilities - can also influence the diffusion process. Sometimes, this last effect is called "mobility effect" to distinguish from the "gradient effect" treated above. Stephenson [80] gave a set of general, coupled equations for the description of atomic current, the resultant stress development and stress relaxation by Newtonian flow (creep). For example the rate of change of the (isotropic) stress (σ_{jj}=-p) in a binary diffusion couple is given by the equation

$$Dp/Dt = -(\Omega_1 \mathbf{divJ_1} - \Omega_2 \mathbf{divJ_2} + 3p/4\eta)[2E/9(1-\nu)]. \tag{3.19}$$

It can be seen that if the creep rate is large ($\eta \Rightarrow 0$), one will arrive at the Darken limit (where the diffusion is controlled by the faster component: $D_D=c_1D_2+c_2D_1$), because the stress relaxation is very fast and lead by the last term in (3.18). If the first two terms can be neglected, an exponential decay of p takes place with a relaxation time $\tau=6(1-\nu)\eta/E$. On the other hand if η goes to infinity the stress field developed will make the difference of the atomic currents equal to zero (feeding back effects of stress gradient in the expression of the atomic currents according to (3.15)) and the interdiffusion coefficient will be determined by the slower component (Nernst-Planck limit) [83]:

$$D_{NP} = D_1D_2/(c_1D_1+c_2D_2). \tag{3.20}$$

For intermediate values of η, depending on the ratio of the diffusion time and τ, different cases are possible. At very short times (small length scales) the atomic currents can be very large (because of the large gradients) and the plastic deformation can be the rate-limiting step for interdiffusion. For an illustration, Fig. 3.8 shows the normalized stress distribution in a planar diffusion couple as calculated in [81]. It can be seen that the stress (or pressure) gradient will have different signs in the center of the zone and outside of it for short times. Accordingly, it can be shown [80,95] that different limits are expected and one can make a classification as follows

a) if	$t < \tau_Q$	$D_{eff} \geq D_D$;	$v \cong v_m/3$
b) if	$\tau_Q < t < \tau_p$	$D = D_{NP}$, $v=0$	
c) if	$\tau_p \cong t$	D_{eff} is determined by plastic flow	
d) if	$\tau_p < t$	$D_{eff} = D_D$, $v=v_m$	

where τ_Q is the time necessary to reach a quasi-equilibrium (i.e to reach the saturation of the stress curves in Fig. 3.8) and v_m is the maximal value of the Kirkendall velocity, v.

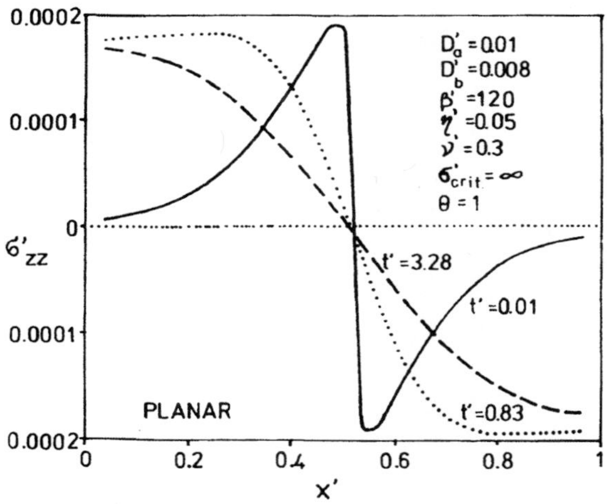

Figure 3.8 Stress versus distance at different times (see also [81])

It is important that usual behaviour is expected only in Nernst-Planck and Darken regimes and strong deviations from this are expected when the process is rate limited (c regime) or if there are transitions between the above regimes. Thus e.g. in these cases, the well-known Boltzmann transformation (introducing a new variable $x/t^{1/2}$) can not be used to get a single curve plotting the composition profiles obtained at different times and the time dependence of the Kirkendall shift will not be parabolic as well.

It can be seen from the above analysis, that although depending on the value of the input parameters (η, which depends also on the temperature, E, initial and boundary conditions, and the diffusion time), deviations from the "normal" behaviour are expected typically on the nanoscale. For example it was shown in [96,97], using the Stephenson model, that the diffusional intermixing in amorphous Zr-Ni multilayers is considerably influenced by the stress gradient built in the diffusion process. This led to on upward curvature on the decay curves of intensity of the small angle X-ray satellites (i.e. a slowing down was observed with increasing time). This stabilization effect - in accordance with the model - was dependent on the modulation length, λ, (Fig. 3.9) and a gradual shift from the Nernst-Planck limit towards the Darken limit was observed with increasing λ.

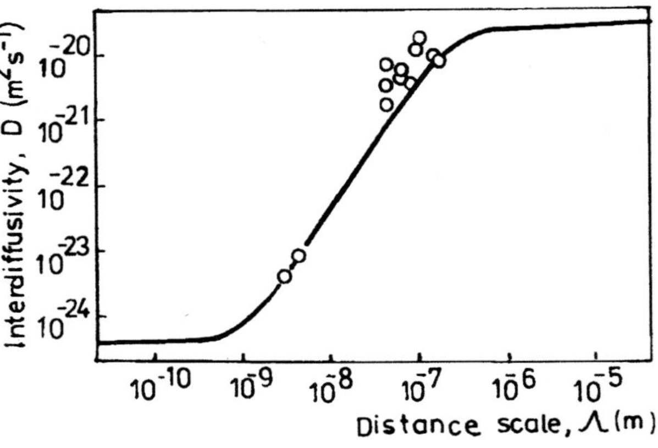

Figure 3.9 D_{eff} as the function of the modulation length in amorphous NiZr multilayers at 529 K [96].

There is a less clear picture for the stress effects in amorphous Si-Ge multilayers. As it was mentioned above, Spaepen [93] concluded from a Cahn-type treatment that stresses can not be important. On the other hand, Fig. 3.10 shows the results of numerical simulations obtained in [82]. It can be seen that, while the stress peak developed on the Ge side relaxes very fast, the large tensile peak on the Si side - due to the fact that the diffusion and creep on this side are much slower - is still present just behind the interface. This sharp peak can explain the strong porosity formation and also its suppression by application of a small hydrostatic pressure as it was observed in [98]. It is also interesting to note (see also [99]) that surprisingly there are no classical bulk interdiffusion measurements in the crystalline Si/Ge system, which can be a consequence of the shift of the sharp interface instead getting a flattening concentration profile, as well as of the possible mechanical failure of such diffusion couples by the sharp tensile stress. For the illustration of the combined effects of stresses and non-linearity, Fig. 3.11 shows the $\ln I/I_0$ versus t curves, which present a significant upward curvature at the beginning. This is obviously the effect of the non-linearity [36] (see also the upper curve for constant intrinsic diffusion coefficients), while stress effects are always reflected in a small slowing down of the intermixing (decrease of the slope). The slopes of curves with "no stresses" are always larger (by about a factor of 2) and stress effect did not cause a considerable curvature in this case.

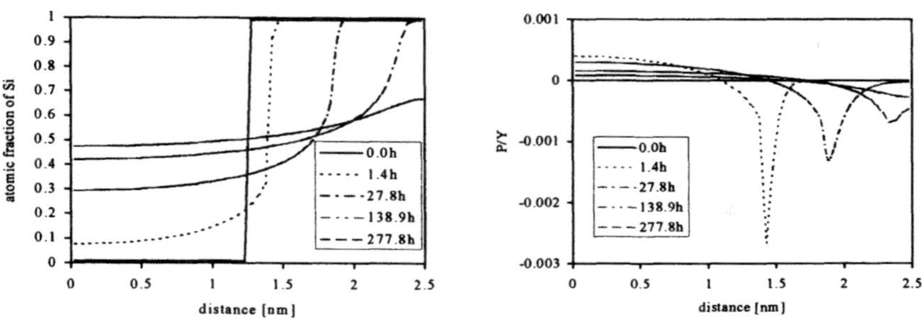

Figure 3.10 Calculated concentration (a) and pressure (b) distributions in amorphous Si-Ge multilayers with 5.8 nm modulation length at different times.

Finally it is worth to mention that in the Cahn-Hilliard treatment - in accordance with the supposition that $j_1 = j_2$ - only the difference of the chemical potentials of the two components had to be determined [74]. Thus, if the gradient energy coefficient κ is not zero, for the use of Stephenson's description a generalisation was necessary. In [82] this generalization was made and thus in principle the investigation of the interplay of the stresses and non-linearity became possible in the framework of the continuum model.

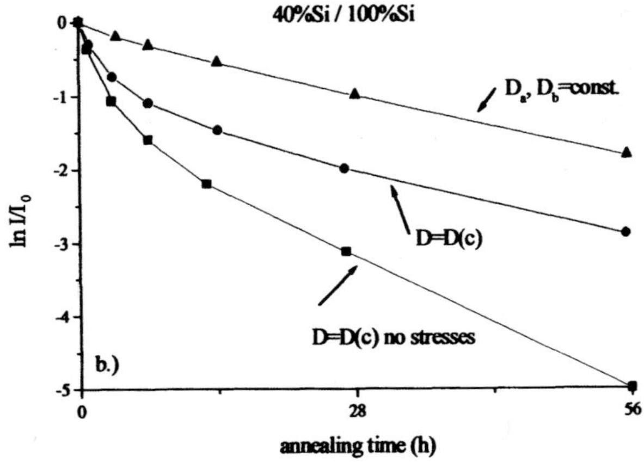

Figure 3.11. Decay curves of Si/$Si_{0.4}Ge_{0.6}$ multilayers with 5.8 nm modulation length

3.4. Segregation kinetics

Usually the segregation is a process during which the redistribution takes place along distances of few nanometers. Furthermore, besides of these kinetic effects, there is also a size effect manifested in changes of the character of the segregation isotherm itself. This is the consequence of the finite size of the sample: either the surface perturbed regions can overlap and thus an increase in the solubility limit can be observed [100,101], or the number of sites at interfaces can be higher than the number of solutes in the sample and a change form the McLean- to a Fowler-Guggenheim-type isotherm can be observed [37]. Related to nanodiffusion, some interesting size effects on the kinetics of surface segregation deserve attention [101, 102, 103] here.

It is usual to classify the binary alloys into two groups: strongly and weakly segregating systems. From the equation of the surface segregation isotherm it is easy to show [104] that there exists a certain bulk composition

$$c_h = \tfrac{1}{2} + (V_{AA} - V_{BB})/4V, \qquad (3.21)$$

so that for bulk compositions of B atoms, c_b, smaller than c_h there is a segregation and for $c_b > c_h$ there is a desegregation. Accordingly, if $c_h < 1$ the system is weakly segregating, while in the opposite case we speak about a strongly segregating system (because now only segregation of B atoms occurs at any bulk composition: $0 \le c_b \le 1$). Furthermore for weakly segregating systems if $T < T_h$ (the temperature of the miscibility gap at c_h), depending on the bulk composition, either segregation or desegregation is expected.

In [102] the interplay between the surface segregation and spinodal decomposition as well as the effect of the slab thickness *in a strongly segregating* spinodal system (with parameters (V_{BB}-V_{AA})=0.15eV and V=V_{AB}–(V_{AA}+V_{BB})/2 = +0.035eV) has been investigated. Figure 3.12 shows the time evolution of the system for three different slab thicknesses. The time shown in the figure can be converted to "real" time by the relation t_{real} / t = exp {-z(V_{AB}+V_{BB})/kT} (see also [105]). It can be seen that – since the system is within the spinodal curve – the kinetics of phase separation is a result of the interplay between the surface segregation, spinodal decomposition and phase separation into equilibrium phases. If the number of layers is large (N=61), the process starts by a fast segregation and a formation of a depleted zone just below the surface. Later on concentration waves develop (the wave length of them increases with time towards that of the optimal spinodal wave) into the bulk and finally a lamellar structure forms with three A-rich and A-poor phases, the concentrations of which correspond to the equilibrium phases. Note that the state at longest computing time for N=61 is still not the final state of the system, since much a longer time would be necessary to reach the equilibrium by coarsening (driven by interface energies only). On the other hand for N=21 and N=13, the size effect can be clearly seen. The system evolves to final states different for the two slabs: for N=13 no spinodal decomposition takes place, but a solid solution is stabilized with surface segregated layers.

The results for a *weakly segregating system* are even amore interesting ((111) direction, z_v=3, z_l=6 at 1000 K with V_{BB}=-0.1555eV, V_{AA}=-0.1728eV and V_{AB}=-0.12975eV) [102]. In this case c_h=0.69 and – from a regular solution model – T_h=2043 K i.e. T<T_h. Figure 3.13 shows that starting from a homogeneous fifty-fifty alloy with 19 atomic planes, first a segregation of B atoms starts at the surface and the segregation generated spinodal wave reaches the centre and overlaps with its symmetrical counterpart. However, there is a competition between the surface segregation of B atoms and the spinodal decomposition and/or the coarsening of the spinodal structure developed. Indeed as it can be seen in this figure, before the B-rich phase could develop at the surface the amplitude of the concentration wave in the centre reached already the concentrations of the equilibrium phases and a relatively stable spinodally decomposed structure has been formed there. Later on the system decreases its free energy by eliminating the first interfaces just below the free surfaces (coarsening of the structure) and as a result *the surface structure makes a sudden "reversal"*. This means that a morphology A/B...B/A develops with a small segregation of B atoms on the surfaces of the A-rich phase. Although the surface energy of such a configuration should be larger than that of a B-rich phase with a small desegregation of B atoms, the overall energy of the system is decreased because of the gain due to the disappearance of two internal interfaces. Furthermore, for the illustration of the effect of the slab thickness, d, Fig. 3.14 shows the situation for 7 atomic planes obtained from the generalization of the Martin's exchange model for a vacancy mechanism. It can be seen that the system arrives at a B/A/B morphology with B rich segregated layers. This is so because, due to the small slab thickness, there are not enough B atoms in the centre to supply the growth of the spinodal wave up to the level of the miscibility gap, and thus the formation of the B-rich phase is not possible there. Thus there exists a critical thickness, d_c, below which the "reversal" of the surface structure does not take place. In [102] it was also shown that the results were the same independently whether the calculations were carried out in the framework of the exchange model or by the generalization of Martin's model for a vacancy mechanism.

It is worth to note that the above type of behaviour is typical only in weakly segregating systems. Indeed in Fig. 3.12, for the strongly segregating system, the surface configuration is very stable and there is no way for the reversal of the configuration. It was also shown in [101] that in this case - if the spinodal wave became "stronger" - there was only a breaking point between the spinodal wave and the surface profile developed at a very early stage of the process. The observed reversal thus is an inherent property of the weakly segregating systems, and the characteristic time evolution of the surface concentration (first there is an increase and then there is a fast decrease of the B atoms at the surface, as well as the existence of a critical slab thickness, d_c, below which this is not observed anymore) seems to be observable experimentally in thin film samples.

3.5. Reaction or diffusion control and nucleation problems

At short diffusion distances in interdiffusion, leading to formation of reaction products, the usual parabolic law of diffusion can be violated, if the role of reactions at the interface is taken into account. At short diffusion times, when the gradient of concentration is large, the reaction controlled rate of the mass transfer (characterized by the coefficient, β, having dimension m/s) can be slower then the diffusion current ($J=-D\text{grad}\rho$, where ρ is the concentration). In this case the growth of the reaction layer will be controlled not by a parabolic, but by a linear law according to the formula [64, 106]:

$$x^2 + Ax = Bt, \tag{3.22}$$

where x is the thickness of the reaction layer, A and B are proportional to β and D, respectively (for small x the second term can be neglected).

Greer [107] illustrated these effects by deriving the following equation for the rate of thickening of the β phase (growing between α and γ phases):

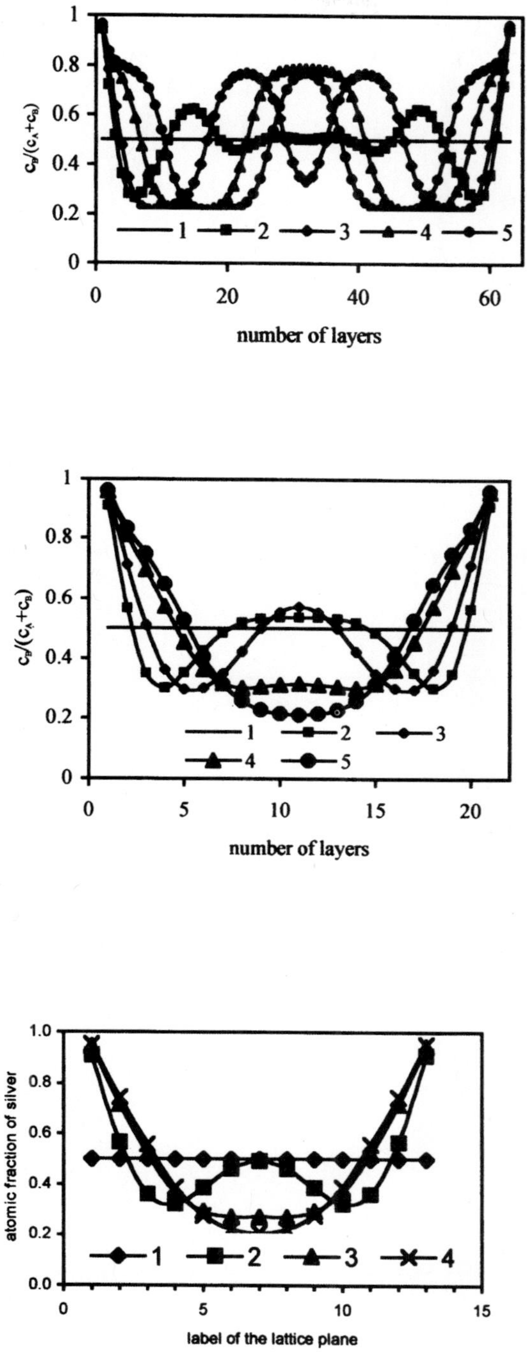

Figure 3.12. Time evolution of composition in a BCC (100) structure for three different slab thicknesses [101]. Increasing numbers at the curves indicate the time sequence.

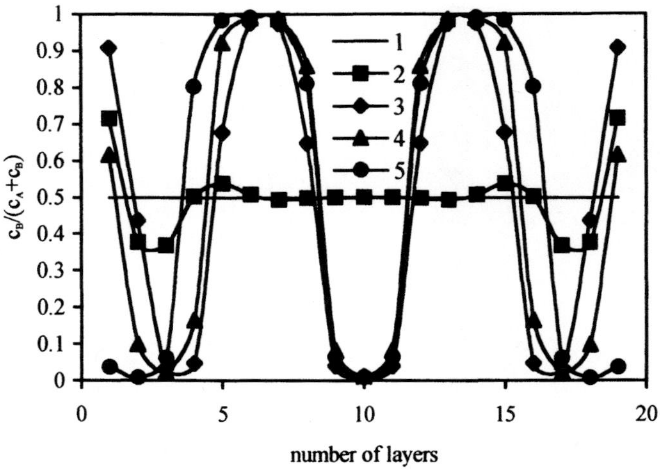

Figure 3.13. Time evolution of composition in a FCC (100) structure with N=19. Increasing numbers at the curves indicate the time sequence.

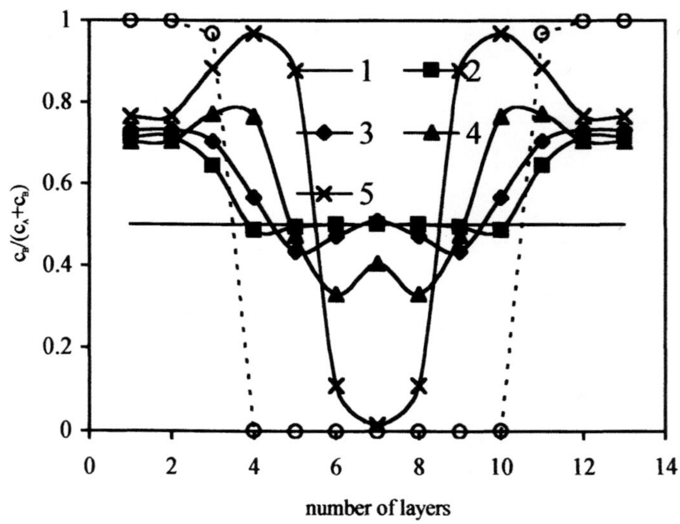

Figure 3.14. The same as Fig. 3.13 but with N=7 and for a vacancy mechanism [102]. The dashed line with open circles denotes the vacancy distribution in the latest state (see also the text). The solid-vapour interface is at the 4th and 10th planes.

$$dx_\beta/dt = H_\beta j_\beta - H_{\beta\alpha} j_\alpha - H_{\beta\gamma} j_\gamma, \qquad (3.23)$$

where e.g. j_β is the flux through the phase β, and coefficients H_β, $H_{\beta\alpha}$ and $H_{\beta\gamma}$ are positive and assumed to be constant through the reaction. It can be seen that there is a competition between the phases growing in a sequence of layers. For example a given phase may not appear (or will dissolve) in the first place if the fluxes in the neighbouring layers are too high. If we are above the limit of the critical thickness, given by the transition between the linear and parabolic growth (see again eq. (3.22), i.e. in macroscopic diffusion couples, the flux through a layer is inversely proportional to its thickness and the thicknesses of the layers can adjust so that each of them survives. In thin films and multilayers however, the thickening may be reaction controlled: then the above relation does not apply (i.e. making the layer thinner the diffusion flux does not increase) and a given phase cannot survive. There are many examples known for such a phase suppression in the thin film case [106, 108]

The same effect can be observed in many bulk or surface coarsening processes (Ostwald ripening [64, 109]) of ensembles of nano-clusters or beaded films on a surface, if their characteristic distance is large enough to produce a diffusion interacting ensemble but shorter than the characteristic length, defined above.

It is also well-known that the nucleation of a new phase (e.g. from a supersaturated solution or from an under-cooled amorphous phase) always needs a certain critical size, d_c, which is typically in the nanometer range. If, for example, in a solid state diffusion reaction the width of the diffusion zone, L, is less than d_c the given phase cannot form. In case of more than one diffusion product, this can lead to different interesting observations in the first stages of the heat-treatments. For example the above competitive growth analysis cannot explain the formation of an amorphous phase - as a first product - in many solid state reactions. For a metastable phase the coefficient H_β is negative and thus this phase should always dissolve in contact with stable phases. Indeed this can be only explained by easy nucleation of these phases. The origin of the nucleation barriers is not very clear, since in many cases the driving force for the interfacial reaction is so large the critical nucleus is in the order of one atom and thus no barrier is expected [110]. Nevertheless, it was illustrated for example in [111] that till the thickness of the diffusion zone is not wide enough to offer the (lateral) critical thickness of the phase to fit into the range between the concentration limits of existence of the phase, it cannot nucleate. Furthermore, because the nucleation should take place in a very steep concentration gradient, this leads to a correction to the Gibbs free energy of nucleus formation; its dependence on the concentration gradient can be also important (this correction is not reduced to the Cahn-Hilliard gradient terms, which were discussed in the previous sections). A recent review of the subject can be found in [112].

ACKNOWLEDGMENTS

The authors are indebted to Pr. G. Erdelyi, Pr. I. Szabo, Pr. E. Moya, Pr. A. Rolland, Dr. Cs. Cserbati, Dr. Ch. Girardeaux, Dr. P. Gas and the coworkers of their groups, particularly Zs. Tokei and Z. Erdelyi, for fruitful collaboration. The work was financially supported by the "Centre National de la Recherche Scientifique" and the OTKA Board of the Hungarian Academy of Sciences.

REFERENCES

[1] Kaur I., Mishin Y. and Gust W., Fundamentals of Grain and Interphase Boundary Diffusion, London, J. Wiley & Sons, 1995.
[2] Suzuoka T., Trans. Japan. Inst. Metals, 1961, 2, 25
[3] Whipple R. T. O., Phil. Mag., 1954, 45, 1225
[4] Le Claire A.D., Brit. J. Appl. Phys., 1963, 44, 351
[5] Levine H. S. and Mc Callum C. J., J. Appl. Phys., 1960, 31; 595
[6] Kaur I., Gust W. and. Kozma L, Handbook of Grain and Interphase Boundary Diffusion Data, Vol 1 and 2, Stuttgart , Ziegler Press, 1989
[7] Beke D.L. General Introduction in "Diffusion in Semiconductors and Non-Metallic Solids", Landolt-Bornstein, New Series, Berlin, Vol 33-B1, ed. Beke D.L., 1998, p 1
[8] Harrison L. G. , Trans. Faraday Soc., 1961, 57, 1191
[9] Hart E. W. , Acta Metall., 1957, 5, 597
[10] Erdelyi Z, Beke D. L., Nemes P., Langer G. A., Phil. Mag. A, 1999, 79, 1757
[11] Beke D. L., Dumas A., Csik A., Langer G. A., Kis-Varga M., Daroczi L., Erdelyi Z., Functional Materials, 1999, 6, 539
[12] Greer A. L., Beke D. L., to be published J. of Non-Equilibrium Processing
[13] E. Emeric, C. Bergman, G. Clugnet, P. Gas and M. Audier, Phil. Mag. Lett., 1998, 78, 77
[14] Belova I. V., Murch G. E., Defect and Diffusion Forum, 2001, 194-199, 1223
[15] Mishin Yu. M. and Razumovskii IM. , Scripta Metall., 1991 25, 1375
[16] Mishin Yu. M. and Razumovskii IM. , Acta Metall. Mater., 1992, 40, 597
[17] Mishin Yu. and Herzig Ch. , Nanostructured Materials, 1995, 6, 859
[18] Szabo I. A., Beke D. L., Kedves F. J., Phil. Mag., A, 1990, 62, 227
[19] Gas P. , Beke D.L., Bernardini J., Phil. Mag., Lett., 1992, 65, 133
[20] Cahn J. W. and Balluffi R. W. , Scripta Metall., 1979, 13, 499
[21] Mishin Yu. M. and Razumovskii IM. , Acta Metall. Mater., 1992, 40, 839
[22] Marvin V. B. , Scripta Metall. Mater., 1995, 33, 901
[23] Glaeser A. M. and Evans, J. M. , Acta Metall., 1986 34, 1545
[24] Fisher J. C. , J. Appl. Phys., 1951, 22, 74
[25] Guthoff F. , Mishin Yu. and Herzig Ch., Z. Metallk., 1993, 84, 584
[26] Koppers M. , Mishin Yu. and Herzig Ch. , Acta Metall. Mater., 1994, 42, 2859

[27] Mishin Yu., Herzig Ch. , Bernardini J. , Gust W., Int. Materials Reviews, 1997, 42, 155
[28] Bernardini J. and Gas P. Defect and Diffusion Forum, 1993, 95-98, 943.
[29] Bernardini J., J. Phys. IV, 1999, 9, 155
[30] Gibbs G. B., Phys. Stat. Sol., 1966, 16, K27
[31] Martin G. and Perraillon B., Grain Boundary Structure and Kinetics, (R. W. Balluffi ed), Amer. Soc. Metals Park, OH 239 1980
[32] Bokstein B. S., Fradkov V. E. and Beke D. L. , Phil. Mag. A, 1992, 65, 277
[33] Mishin Yu. and Herzig Ch. , J Appl. Phys., 1993, 76, 8206
[34] Bernardini J., Tokei Zs. and Beke D. L., Phil. Mag., A, 1996, 73, 237
[35] Bokstein B. S., Ostrovksy A. S. and Rodin A. O. , Phil. Mag. A, 1995, 72, 829
[36] Beke D. L., Godeny I., Erdelyi G. and Kedves F. J., Phil. Mag., A, 18987, 56, 659
[37] Cserhati Cs., Szabo I.A., Beke D.L., J. of Appl. Phys., 1998, 83, 3021
[38] Ostrovsky A. S., Bokstein B. S., Appl. Surf. Sci., 2001, 173, 312
[39] Balandina, N., Bokstein, B., Ostrovsky, A., Def. and Dif. Forum, 1996, 129-130, 151
[40] Balandina, N., Bokstein, B., Ostrovsky, A, Def. and Dif. Forum, 1997, 143-147, 1499
[41] Wildman H. S., Howard J. K. and Ho P. S. , J. Vac. Sci. Technol., 1975, 12, 75
[42] Hwang J. C. M., Pan J. D. and Balluffi R. W., J. Appl. Phys., 1979, 50, 1339
[43] Erdelyi Z., Girardeaux Ch., Langer G. A., Beke D. L., Rolland A. and Bernardini J., J. Appl. Phys., 2001,89, 3971
[44] Gleiter H., Phys. Stat. Sol.b, 1992, 172, 41
[45] Gaffet E., Begin-Colin S. and Tillement O., in "Matériaux", Innovation 128, Tech. Tendances,S. A. Paris, 1998
[46] Klotsman S. M., Defect Diff. Forum, 1993, 99-100, 25
[47] Siegel R W, in Materials Interfaces:, D. Wolf and S. Yip Eds., p 431, London , Chapman and Hall, 1992 .
[48] Beke D. L., Defect and Diffusion Forum, Transtec. Public. Ltd, 1995, 103, 51
[49] Hofler H. J., Hahn H. and Averback R.S, Defect and Diffusion Forum, 1991, 75, 195
[50] Valiev R.Z., Islamgaliev R.K. and Aleksandrov I.V. , Prog. Mater. Sci.,2000, 45, 102
[51] Kolobov Yu. R., Grabovotskaya G.P., Ivanov M.B., Valiev R.Z. and Love T.C., in "Investigations and Applications of Severe Plastic Deformation", NATO Science Series 80, ed. T. C. Love and R.Z. Valiev, Dordrecht, Kluwer Academic Publishers, 2000, p 161
[52] Kolobov Yu. R., Grabovotskaya G.P., Ivanov M.B. , Zhilyaev AA.P. and Valiev R.Z., Scripta Mater., 2001, 44, 873
[53] Fecht H.J., Hellstern E., Fu Z. and Johnson W.L. , Metall. Trans. A, 1990, 21, 2333
[54] Fultz B. and Frase H.N., Hyperfine Interactions, 2000, 130, 81
[55] Eastman J. A., Fitzsimmons M. R., Thompson L. J., Lawson A. S., Robinson R. A., Nanostructured Mater.,1992, 1, 465,
[56] Huang Y. K., Menovsky A. A. and de Boer F. R., Nanostructured Mater., 1993, 2, 587
[57] Kuwano H., Ouyang H. and Fultz B., Nanostruct. Mater., 1992, 1, 143
[58] Fultz B., Kuwano H. and Ouyang H., J. Appl. Phys., 1995, 77, 3458
[59] Larikov L., Sidorenko S., Deyneka N. and Litvinova, Mem. Noe.Mex., 1999, 2, 21
[60] Losch W and Jardim P. M., Scripta Mater., 1998, 38, 1857
[61] Fujita T., Horita Z. and Langdon T. G., Phil.Mag., A, in press
[62] Tanimoto H., Pasquini L., Prummer R., Kronmuller H., Schaefer H. E., Scripta Mater., 2000, 42, 961
[63] Wurschum R., Reimann K., GruB S., Kubler A, Scharwaechter P., Frank W;, Kruse O., Carstanjen H. D., and Schaeffer H. E., Phil.Mag B, 1997, 76, 407
[64] Philibert, J., Atom Movements. Diffusion and Mass Transport in Solids, Les Ulis, France, Les Editions des Physique, Paris, 1991
[65] Martin, G. and Benoist, P., Scripta Met, 1977, 11, 503
[66] Martin, G., Phys. Rev., B, 1990, 41, 2297
[67] Erdélyi, Z., Beke D.L., Nemes, P. and Langer G.A., Phil. Mag. A, 1999, 79, 1757

[68] Cserháti Cs., Bakker, H. and Beke, D.L., Surf. Sci. , 1993, 290, 345
[69] Erdélyi, Z., Beke D.L., and Cserháti, Cs., to be published
[70] Csik A., Langer, A.G., Beke, D.L., Erdélyi, Z., Menyhárd, M. and Sulyok A., J. of Appl. Phys., 2001, 89, 804
[71] Erdélyi, Z., Girardeoux, Ch., T□kei, Zs., Beke, D.L., Cserháti, Cs. and Roland A., Surf. Sci. in press
[72] Martin, G. in "Diffusion in Materiials"(eds. A.L. Laskar J. C. Bocquet, G. Brebec, C. Monty) Nato ASI Series Series E Applied Sciences, Vol. **179.** ,Kluwer Academic Press, Dordrecht, 1990, p. 129
[73] Greer, A.L. and Speapen, F., in *"Synthetic Modulated Structures"* (Eds. Chang, L.L. and Giessen, B.C.), Academic Press, New York, 1985. p. 419
[74] Cahn, J.W. and Hilliard, J.E., J Chem Phys., 1958, 28 258
[75] Hillert, M. Acta Metall., 1961, 9, 29
[76] Yamauchi, H. and Hilliard, J.E., Scipta Metall, 1972, 6, (909
[77] Szabó, I.A. and Beke. D.L. to be published
[78] Cook, H.E. and deFontaine, D., Acta Metall., 1969, 17, 915
[79] Cook, H.E. and deFontaine, D., Acta Metall., 1971, 19, 607
[80] Stephenson, G.B., Acta Metall., 1988, 36, 2663
[81] Daruka, I., Szabó, I.A., Beke, D.L. Cserháti, Cs., Kodentsov, A. and Van Lkoo, F.J.J., Acta Metall., 1996, 44, 4981
[82] Beke, D.L., Nemes, P., Erdélyi, Z., Szabó, I.A. and Langer, D.G., Mat. Res. Soc. Symp. Proc. 1998, 527, 99
[83] Beke. D.L. in "Diffusion and stresses" (eds. Beked, D.L. Szabó, I.A.), Trans Tech. Publ., Zurich, 1995, p. 9
[84] Beke, D.L. Opposits, G. and Szabó, I.A., in "Stress Induced Phenomena in Metallization; Fifth International Workshop, (eds. O. Kraft, et.al.) AIP Conference Proc. , 1999, 491, 77
[85] Darken, L.S. Trans Am. Inst. Min. Met. Eng., 1948, 175, 184
[86] Schmalzried, H., "Chemical Kinetics of Solids", VCH Verlagsgesellshaft, Weinheim, 1995, p.336
[87] Herring, C., J. Appl. Phys. , 1950, 21, 437
[88] Cahn, J.W., Acta Metall., 1961, 9, 795
[89] Cahn, J.W., Acta Metall., 1962, 10, 179
[90] Larche F.C. and Cahn, J.W., Acta Metall., 1985, 33, 331
[91] Larche, F.C., Woorhees, Defect and Diffusion Forum, 1996,129-130, 31
[92] Baranowsky, B., J. Less-Common Met., 1989, 154, 329
[93] Spaepen, F., J. Magn. Magn. Mat., 1996, 156, 401
[94] Geguzin, Yu. E. and Krivoglaz, M.A., "Studies in Soviet Science, Migration of macroscopic inclusions in solids", Consultant Bureau, New York, 1973, p.75
[95] Beke, D.L. and Szabó, I.A., Defect and Diffusion Forum, 1993, 95-98, 537
[96] Bottiger, J., Greer, A.L. and Karpe, N., Mater. Sci. Eng., A, 11994, 78, 65
[97] Greer, A.L., Defect and Diffusion Forum, 1966, 66-69, 163
[98] Daróczi, L., Beke, D.L., Langer, G.A., Radnóczi, Gy. and Czigány, Zs., J of Magn. Magn. Mat., 1996, 156, 417
[99] Beke D.L., langer, G.A., Csik, A., Erdélyi, Kis-Varga, M and Szabó, I.A., Papp, Z., Defect and Diffusion Forum, 2001, 194-199, 1403
[100] Beke, D.L., Cserháti, Cs. and Szabó, I.A., Nanostr. Mat., 1997, 9, 665
[101] Beke, D.L., Erdélyi, Z., Bakos, P. Cserháti, Cs. And Szabó I.A., Porc. of Int. Conf. on Soild-Solid Phase Transformations, 99 (JIMIC-3))eds. Koiwa, M., Otsuka, K, Miyazaki, T.) The Jap. Inst. of Metals, 1999, p. 1297
[102] Erdélyi, Z., Bakker, H. and Beke D.L., to be published

[103] Vasyljev, M.O. and Sidorenko, S.I., „Diffusion and surface segregation", Ministry of Education of Ukraine, Series of Mongraph „Problem sof Applied and Theoretical Material Science" , Vol. 1. Kiev, 1998
[104] Cserháti, Cs., Szabó, I.A. and Beke, D.L., Nanostr. Mat., 1998, 10, 195
[105] Cserháti, Cs., Bakker, H. and Beke D.L., Surf. Sci., 1993, 290, 345
[106] U. Gössele, K.N. Tu, J. Appl. Phys., 1982, 53, 3252
[107] Greer, A.L. in Comprehensive Composite Materials (Eds. Kelly A. and Zweben C.) Vol.3 Metall-Matrix Composites (ed. T.W. Clyne)
[108] Gas, P., d'Heurle, F.M., Diffusion in Silicides in "Diffusion in Semiconductors" Landolt-Börnstein, New Series, Vol. 33-A (ed. Beke D.L.) Berlin, 1998
[109] Beke, D.L. and Kaganovskii, Yu. S., Mat. Sci. and Engineering, B, 1995, 32, 185
[110] d'Heurle, F.M., J. Mater. Res., 1988, 3, 167
[111] Desre, P. and Yavari, A.P., Phys. Rev. Letters, 1990, 64, 1553
[112] Gusak, A.M., Hodaj, F. and Bogatyev, A.O., J. Phys.:Condens. Matter, 2001, 13, 1

NANOIONICS AND SOFT MATERIALS SCIENCE

Joachim Maier
Max-Planck-Institut für Festkörperforschung, 70569 Stuttgart, Germany
s.weiglein@fkf.mpg.de

1. INTRODUCTION

Point defects (which include electronic carriers in this context) are the relevant charge carriers in solids. Their concentrations and mobilities determine mass transport and charge transport, and are thus key parameters as regards chemical and electrochemical kinetics.

They sensitively influence and often determine the kinetics of preparation, processing, conditioning and degradation of ceramic materials. As far as electroceramics such as battery electrodes, high temperature fuel cells or ceramic electrochemical sensors are concerned, they are decisive for the function and performance. The significance of nano-structured materials in this respect is twofold: (i) The ratio of interfacial area to volume is increased, such that the properties of many materials become interfacially controlled. As long as the local relevant parameters are still those of the massive crystal we may speak of "trivial size effects" [1]. (ii) "True size effects" occur if the reduced spacing of interfaces leads to local effects.

In the first part of the paper bulk defect chemistry is briefly described which defines the equilibrium concentration of the ionic and electronic point defects. In the second part modifications due to the presence of interfaces are treated first for semi-infinite and then for finite boundary conditions. Not only are local properties considered but also their superposition to the overall response of the material. Then in the central part it is investigated how concentrations and conductivities depend on the spacing of interfaces. In other words this third part describes size effects being crucial for nano-ionics. A feature of special interest is the phenomenon of interfacial phase transitions which, however, can only be touched upon. Finally, the potential of nano-ionics is explored together with the trend to a "soft materials science" exhibiting a high compositional, structural, functional complexity on different hierarchic levels.

Recently various reviewing articles on the topic of "nanoionics" [2, 3, 4, 5, 6, 7, 8, 9, 10, 1] became available. Some of them [3, 4, 5] address the significance of the nano-regime for batteries, photo-electrochemistry and related areas being well-suited for complementary information. A good overview on electronic effects is given in Ref. [11].

2. POINT DEFECTS AND CONDUCTIVITY IN THE BULK

Point defects are chemical excitations of the perfect ground structure and play the same role in the solids, as H_3O^+ and OH^- do in water [12, 13, 14]. Fig. 1 illustrates the correspondence of exciting valence

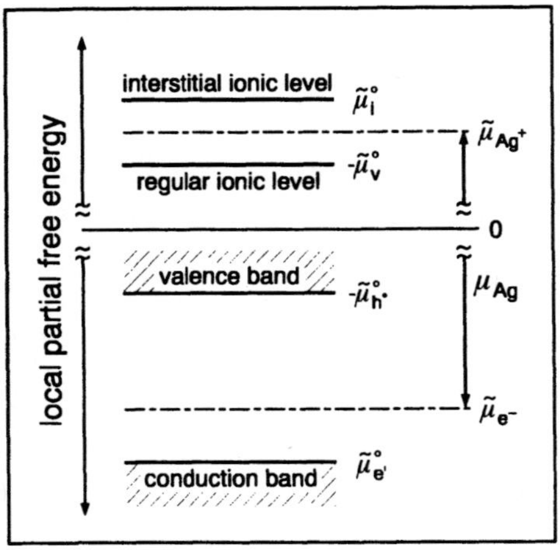

Figure 1. Ionic and electronic disorder for AgCl in terms of level diagrams. The electronic and ionic "Fermi-levels" are connected via the chemical potential of Ag, and thus given by the exact position in the phase diagram. The ionic disorder is much higher than the electronic. The prevailing of conduction electrons over holes corresponds to a slight Ag excess in AgCl [8].

electrons into the conduction band (formation of e′) connected with the formation of holes in the valence band (h˙), with the exciting of regular silver ions in silver chloride into interstitial positions (formation of interstitial silver ions, i. e. $Ag_i^{\cdot}$) connected with the formation of vacancies (i. e. of V'_{Ag}). Similarly, H_3O^+ and OH^- can be considered as the ionic point defects in the liquid of water phase. The figure does not only reveal that the ionic and electronic point defects are the acid-base and redox active particles which constitute the difference between the real structure and the perfect structure, it also shows that the two "Fermi-levels" (i. e. the two electrochemical potentials) $\tilde{\mu}_{Ag^+}$ and $\tilde{\mu}_{e^-}$ are related via the chemical component potential (in silver halides: $\mu_{Ag} = \tilde{\mu}_{Ag^+} + \tilde{\mu}_{e^-}$) and thus by the exact position in the phase diagram (exact silver-to-halogen ratio). In addition to the Frenkel disorder reaction shown, (simple binary) solids can — unlike fluids — exhibit four different intrinsic ionic disorder reactions, viz. to Frenkel disorder, Schottky disorder, Anti-Frenkel and Anti-Schottky disorder (see Table 1). (We neglect anti-site disorder.) For a binary solid M^+X^- one out of the four reactions is redundant. The list of the defect chemical reactions in a pure material is completed

Table 1. Basic intrinsic ionic disorder reactions in a crystal M^+X^-. In the Kröger-Vink notation M_M denotes a M^+ ion on a M^+-site, X_X a X^- ion on a X^--site. V_M': vacant M^+ site (which has the effective negative charge -1), $V_X^{\cdot}$: vacant X^- site (effective charge +1); V_i: vacant interstitial site; $M_X^{\cdot\cdot}$ and X_M'' denote the anti-site defects, viz. M^+ on X^--site and X^- on M^+ site respectively. The anti-site disorder* is only realistic in materials with largely covalent bonds (GaAs), i. e. charges are formal charges. Site substitution is more favorable in multinary compounds, viz. site exchange between two cation or two anion-sublattices (e. g. Y^{3+} and Ba^{2+} in $YBa_2Cu_3O_{6+x}$).

Type of reaction	Description	Reaction in Kröger-Vink notation
Frenkel reaction	excitation of a cation from a regular site to an interstitial site	$M_M + V_i \rightleftharpoons M_i^{\cdot} + V_M'$
Schottky reaction	formation of vacant cation and anion sites by releasing a MX unit	$M_M + X_X \rightleftharpoons MX + V_M' + V_X^{\cdot}$
Anti-Frenkel reaction	excitation of anion from a regular site to an interstitial site	$X_X + V_i + MX \rightleftharpoons X_i' + V_X^{\cdot}$
Anti-Schottky reaction	formation of interstitial defects by incorporation of additional MX unit	$MX + 2V_i \rightleftharpoons M_i^{\cdot} + X_i'$
(Anti-site reaction*)	(site exchange between cation and anion)	$(M_M + X_X \rightleftharpoons M_X^{\cdot\cdot} + X_M'')$

by the interaction reaction which the neighboring phase that defines the exact stoichiometry. The missing equation for solving the reaction scheme in terms of the defect concentrations is — as long as we refer to the bulk — the electroneutrality equation. Fig. 2 displays the situation for the Frenkel disordered AgCl. Since electronic carriers do not play a major role in the electroneutrality equation (primarily ionic disorder $[V_{Ag}'] \simeq [Ag_i^{\cdot}]$ (see Fig. 2, top)) the ionic defect concentration immediately follows from the Frenkel reaction for which a mass action law is valid. (The validity of the mass action follows from the minimization of the free enthalpy, which is not just proportional to the number of defects but also contains a non-linear configurational entropy contribution.) Owing to their high concentration, the ionic defect concentrations

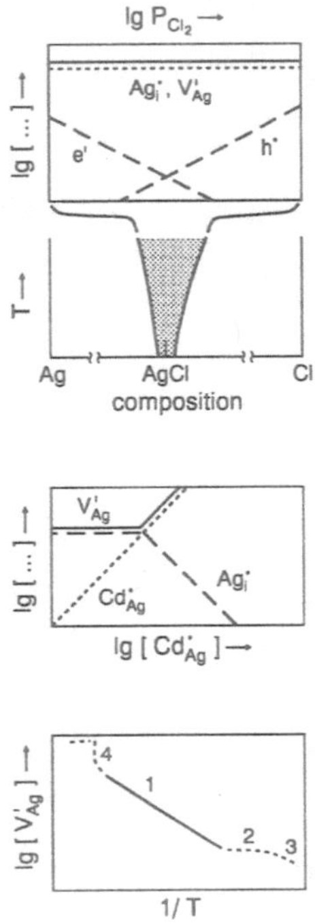

Figure 2. Defect chemistry in the bulk of AgCl [15].

are not significantly affected by the position in the phase diagram. (Note that this only holds true for the relative change which matters in the mass action constants and logarithmic representations.) The concentrations of the electronic carriers as minority species, however, depend sensitively on the outer silver or chlorine partial pressure defining the exact stoichiometry (see Fig. 2, top). In AgCl conduction electrons prevail over the holes if AgCl is in contact with Ag while the converse situation is met if AgCl is equilibrated with a pure Cl_2 atmosphere (1 bar). Finally, aliovalent impurities enter the electroneutrality equation and seriously modify the equilibrium picture (see Fig. 2, center). The introduction of positively charged defects (e. g. Cd^{2+} on an Ag^+ site leading to the defect $Cd^{\cdot}_{Ag}$) leads to an increase of all negatively charged defects (V'_{Ag} and e') and a decrease of all positively charged defects ($Ag^{\cdot}_i$ and $h^{\cdot}$). This rule of homogeneous doping (z_k, c_k: charge number and concentration of mobile defect k; z, C: charge number and concentration of the dopant)

$$\frac{z_k \delta c_k}{z \delta C} < 0 \tag{1}$$

follows from the combination of the electroneutrality equation and individual mass action laws [12]. If the electroneutrality equation can be written as a proportionality (which is always the case if two oppositely charged defects are prevailing, and does not necessarily imply ionic disorder), the defect concentration can be written as [12, 16]

$$c_k = \alpha_k(\Pi_p P_p^{N_{pk}})(\Pi_q C_q^{M_{qk}})(\Pi_r K_r^{\gamma_{rk}}(T)). \tag{2}$$

(α_k denotes constants of proportionality stemming from different charge numbers.) We recognize that temperature (T), component partial pressure (P) and doping content (C) are the decisive control parameters to tune the charge carrier budget (defect chemistry) of a given material. Both partial pressure (e. g. P_{Cl_2} over AgCl) and doping content (e. g. $CdCl_2$ in AgCl) enter Eq. (2) similarly. A basic difference between both parameters is that the first can be varied in-situ while the dopants are considered to be immobile and to be introduced ex-situ (usually under preparation conditions). Intrinsic point defects of immobile native constituents (e. g. Sr-vacancies in $SrTiO_3$ at moderate temperatures) also lead to such ex-situ parameters. Frozen impurities (or dopants) usually become mobile at very high temperatures, an actual example being the hydroxide content in water containing oxides [17, 18]. (At high temperatures P_{H_2O} then belongs to the list of P-parameters while at low T the water content enters the list of C-parameters.) Conversely, at sufficiently low temperatures all ionic defects become immobilized (increasing the C-list) while the purely electronic reactions still may be fast enough. A typical situation is Fe-doped $SrTiO_3$ in which around room temperture the vacancy concentration is fixed by the pretreatment while redox changes of the iron impurities are still active. A systematic treatment is given in Ref. [19].

Let us consider the T-dependence: Eq. (2) describes an Arrhenius type of dependence which is realized in Fig. 2 in the intrinsic case (pure material or doped material at very high temperatures) but also in the extrinsic case (dopant governs the electroneutrality condition) at sufficiently high temperatures (see bottom curves 1, 2). Decreasing the temperature increases the Coulomb trapping of the charge carriers. This can be approximately described by association equilibria [20] (which augments the number of variables k but also increases the list of K's) (curve 3). A deviation from the Arrhenius dependence is noticed at high temperatures which can be attributed to a long-range Coulomb interaction. In many cases the resulting activity correction can be much better described by a cube root law ($\mu^{ex} \propto c^{1/3}$) in c [21] than by a Debye-Hückel corection ($c^{1/2}$) [20]: This behavior is immediately explained by a topological order, the energetic effect of which corresponds to superimposing a defect lattice on the perfect lattice. The resulting decrease of the effective formation energy leads to an avalanche (curve 4) and eventually to a phase transformation (superionic state) [22, 23]. Usually this is accompanied by a structural change leading to a sublattice-molten phase of a particular structure (α-AgI) or even to a breakdown of the total lattice (e. g. molten AgCl).

The Fermi-levels shown in Fig. 1 are electrochemical potentials ($\tilde{\mu}$) while the fixed energy levels are electrochemical potentials minus configurational contributions ($\tilde{\mu}^\circ$) [14]. Thus, in the case of interaction they also include the respective energetic contributions. The configurational terms are of Boltzmann type if the defects are diluted, and more complex if concentrated. The simplest correction for the carrier k is a Fermi-Dirac correction RT ln $\frac{n_k}{n-n_k}$ (where n_k denotes the number of defects and n the number of available sites). The standard electrochemical potentials, $\tilde{\mu}^\circ$, can be split into a standard chemical potential and an electrical potential term. While the second term can also vary in equilibrium, the change of the first term requires structural modifications if we restrict to ionic carriers [8]. This is in contrast to electrons which perceive a quantum mechanical confinement effect. Discrepancies between ions and electrons are also reflected in the mobilities (u_k). A non-zero mobility is a prerequisite for a finite conductivity of the point defect. For ionic carriers as well as for polaronic electronic carriers the mobility is activated by an effective migration threshold. The mobility of nearly free electronic carriers, however, is not thermally activated, it rather decreases when T is increased due to increased scattering effects.

3. INTERFACES: CORE AND SPACE CHARGE EFFECTS

In the following we consider the internal interface to consist of a core layer (thickness of the order of interatomic spacing, $\sim |s|$) sandwiched between two space charge layers (thickness of the order of the Debye length λ). While in the core the structure is altered, and so are carrier mobilities, the space charge regions are ideally zones in which the bulk structure is essentially maintained (and so the standard potential μ° and the mobility u), i. e. only an altered point defect density is established. This abrupt interface model is not absolutely realistic, e. g. structural gradients reflected by elastic fields make the junction more continuous, yet, the picture can be taken as a first approximation [24, 25]. The altered core structure is simply a consequence of the different structures or orientations of the two phases in contact. The adjustment between the interfacial core and the bulk structure by the modified defect structure in the space charge regions is thermodynamically expressed by the bending of the standard (partial free) energy level. Since μ° is invariant, this bending is due to the electric field term. The electric potential variation is a consequence of the breakdown of the electroneutrality equation which has to be replaced by Poisson's equation in full generality. Thus the charge density ρ is non-zero and determined by the curvature of the electrical potential ϕ, i. e. $\rho \propto \phi''$.

While in complete equilibrium (Gouy-Chapman layers) the extent of the space charge zones depends on bulk parameters only, the degree of level bending (space charge potential) is additionally characterized by the defect density at the layer (x = 0) which is adjacent to the core layer (x = s). This value definitely depends on the properties of the interface. The prediction of the space charge effect just from structure and composition of bulk and interface or even from the properties of the phases brought into contact is a tremendous challenge for future theoretical and experimental studies. An appropriate thermodynamic framework is given by the core-space charge picture described in Ref. [24]. An appropriate working parameter is the "degree of influence" [25]

$$\vartheta_k = \frac{(c_{k0}/c_{k\infty})^{1/2} - 1}{(c_{k0}/c_{k\infty})^{1/2} + 1}. \tag{3}$$

($c_{k0}, c_{k\infty}$ being the concentrations of k at x=0 and in the bulk, respectively.) It is +1 for the maximum enrichment of the carrier k, -1 for a maximum depletion and 0 for a zero effect. Since all mobile defects perceive the same electrical field, a given interfacial effect (measured by the charge density Σ in the interfacial core) leads to a predictable modification of the bulk densities according to

$$\frac{z_k \delta c_k}{\delta \Sigma} < 0. \tag{4}$$

This relation may be compared to Eq. (1) and will render good service as a "rule of heterogeneous doping" [12] as described below.

This enormous variability of the defect concentrations at interfaces can hardly be overestimated as regards the influence on electrical and electrochemical properties. In this way, conductors can be changed into insulators, ion conductors may be turned into mixed or even electronic conductors and vice versa.

It is crucial to distinguish between two types of profiles. One is realized if both majority carriers k = 1, 2 are mobile. Then Gouy-Chapman profiles are established in equilibrium [26]. For large effects ($|\vartheta| \longrightarrow 1$) they simplify to

$$c_1(x) = \frac{c_{10}}{(1 + \sqrt{c_{10}/c_{\infty}}\, x/2\lambda)^2} \tag{5}$$

for the enriched majority carrier. The concentration of the depleted counter carrier follows from $(c_2/c_{2\infty}) = (c_{1\infty}/c_1)^{|z_2/z_1|}$ with $z_2 c_{2\infty} \simeq$

$z_1 c_{1\infty}$ (z_1, z_2 are the charge numbers). (The fact that for $x \longrightarrow \infty$, c_1 drops below $c_{1\infty}$ in the above equation is a consequence of the approximation and is not of interest since the range of validity is restricted to the range between $x=0$ and $x=2\lambda$ where the bulk value is reached.) If we integrate $c_1(x)$, multiply with mobility and charge of the carrier 1, we obtain the (parallel) space charge conductance (area related) as being characterized by an effective concentration $\sqrt{c_1 c_\infty}$ and an effective thickness 2λ:

$$Y_1^{||} = |z_1| F u_1 (2\lambda) \sqrt{c_{10} c_\infty}. \tag{6}$$

As the Debye-length λ is determined via

$$\lambda = \sqrt{\frac{\varepsilon RT}{2 z_1^2 F^2 c_{1\infty}}}, \tag{7}$$

$\Delta Y^{||}$ becomes $u_1 \sqrt{2\varepsilon RT c_{10}}$ and thus independent of the bulk value. Note that this result is also proportional to the excess charge density in the core.

A second type of profile which is only important for depletion effects, is often met in doped systems and is called Mott-Schottky profile [27]. If the dopant is immobile and the majority counter defects are depleted, the space charge density is essentially determined by the frozen defect concentration, hence x-independent; this leads to a half-Gaussian profile for the depleted carrier 2

$$c_2/c_{2\infty} = \exp - \left|\frac{z_2}{z_1}\right| \left(\frac{x - \lambda^*}{2\lambda}\right)^2. \tag{8}$$

The space charge zone, which also defines the range of validity of Eq. (8), is characterized by λ^* which is, in contrast to λ, now influenced by the space charge potential and thus by the ratio $c_{20}/c_{2\infty}$:

$$\lambda^* = \lambda \sqrt{4 \frac{z_1}{z_2} \ln \frac{c_{20}}{c_{2\infty}}}. \tag{9}$$

The reader should pay attention to the fact that (i) λ^* can be greater than λ and that (ii) λ is now fixed by the impurity content ($c_{1\infty}$). The space charge resistance (times area), $Z^\perp$, is obtained via integration of $1/c_2(x)$ yielding an effective concentration $2c_{20} \ln \frac{c_{2\infty}}{c_{20}}$ and an effective thickness λ^* (u_2: roughly constant)

$$Z^\perp = \frac{\lambda^*}{|z_2| F u_2 2 c_{20} \ln \frac{c_{2\infty}}{c_{20}}}. \tag{10}$$

Table 2. Possible space charge situations for acceptor doped (D′) metal rich oxides with oxygen vacancies and excess electrons as decisive carriers, and corresponding effective conductivity and resistivity expressions for electrons (n) and oxygen vacancies (V), respectively [28].

model	concentration profile	effective electronic conductivity, $\sigma^{\parallel}_{m,n}$ and ionic resistivity, $\rho^{\perp}_{m,V}$
Schottky-Mott	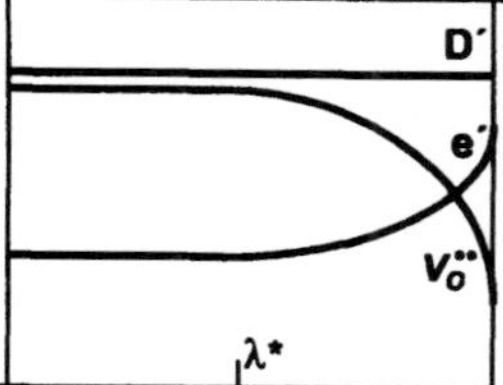	$\sigma^{\parallel}_{m,n} \propto \lambda^* \frac{c_{n0}}{2\ln(c_{n0}/c_{n\infty})}$ $\rho^{\perp}_{m,v} \propto \lambda^* \frac{1}{2c_{v0}\ln(c_{v0}/c_{v\infty})}$
Gouy-Chapman	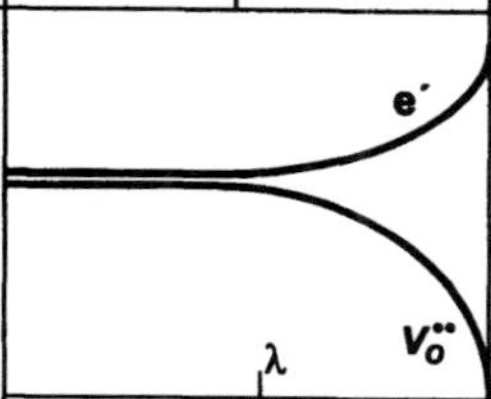	$\sigma^{\parallel}_{m,n} \propto (2\lambda)\sqrt{c_{n0}c_{n\infty}} \propto \sqrt{c_{n0}}$ $\rho^{\perp}_{m,v} \propto (2\lambda)\frac{1}{\sqrt{c_{v0}c_{v\infty}}} \propto \frac{1}{\sqrt{c_{v0}c_{v\infty}}}$
	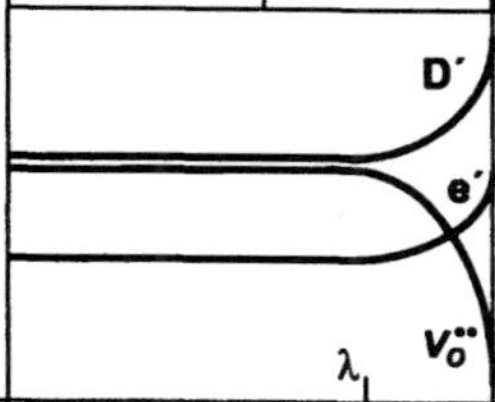	$\sigma^{\parallel}_{m,n} \propto (2\lambda)\sqrt{c_{n0}c_{n\infty}} \propto \frac{1}{\sqrt{c_{v\infty}}}\sqrt{c_{n0}c_{n\infty}}$ $\rho^{\perp}_{m,v} \propto (2\lambda)\frac{1}{\sqrt{c_{v0}c_{v\infty}}} \propto \frac{1}{\sqrt{c_{v0}c_{v\infty}}}$
combined	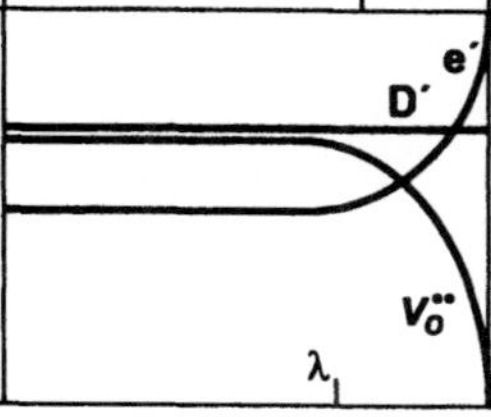	$\sigma^{\parallel}_{m,n} \propto (2\lambda)\sqrt{c_{n0}c_{v\infty}} \propto \sqrt{c_{n0}}$ $\rho^{\perp}_{m,v} \propto (2\lambda)\frac{1}{\sqrt{c_{v0}c_{n\infty}}} \propto \frac{1}{\sqrt{c_{v0}}}\frac{1}{\sqrt{c_{v\infty}c_{n\infty}}}$

∞ 0

A more systematic compilation is given by Table 2 [28].

While the effects on the mobile defects just discussed are thermodynamically required — and the mobilities remain unchanged in the abrupt model — the impact on carrier mobilities, but also on densities within the structurally modified core are more difficult to estimate. In order to visualize this last point we will consider an elemental crystal for which we can ignore ionic space charge effects. Since the defect formation energy in the interfacial core is clearly different from the bulk value owing to altered coordinations and bond strengths and say modified by a factor β, the equilibrium concentration in the core is proportional to the bulk value to the power of β. Let us consider, to give an example, a gas/solid interface, ignore relaxation effects, hence taking $\beta = 2/3$, and assume a bulk concentration of 10 ppm and a molar volume of 30 cm^3/mole, then the core defect concentration is assessed to be greater by a factor of 50 compared to the bulk [1]. If we consider ionic crystals, analogous considerations are only possible if both interfacial core and interior are very disordered (e. g. very low defect formation energies) or structurally very similar. Generally, however, the junction between two structurally different regions automatically leads to a charging and to space charge zones. In addition the mobility is different in the structurally perturbed layer. (In metals it is well known that the mass transport along boundaries can be much enhanced compared to the bulk. In the case of ionic crystals this might also be the case for low mobility materials.)

It is to be expected that for ionic materials the bulk of which is characterized by a low concentration of fast defects, the space charge effects will be of paramount significance. Let us briefly discuss a few examples which prove decisive in the context of nanocrystalline ionic materials. First let us refer to accumulation layers leading to conductivity

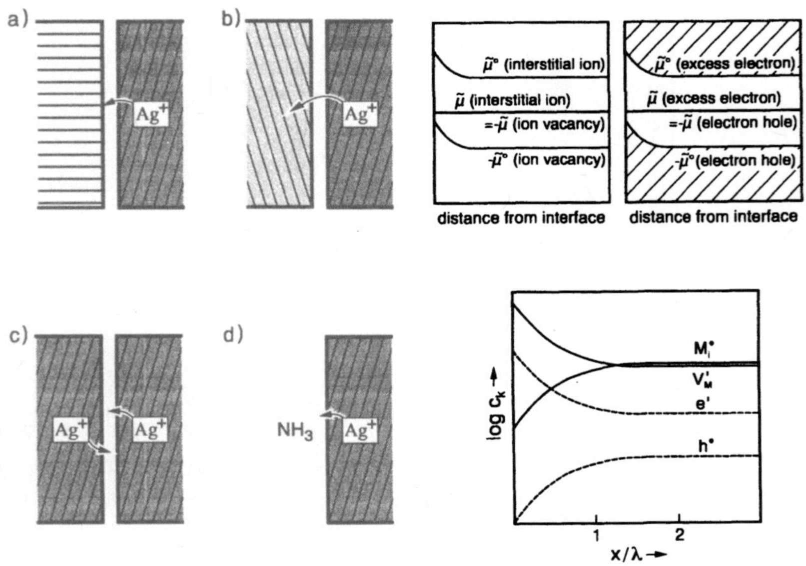

Figure 3. Four basic situations concerning vacancy accumulation layers [29, 7].

enhancement: Four basic situations (see Fig. 3) are worth being considered here [30]: (i) The first (a) refers to the contact of the conductor with a second insulating solid phase which exhibits surface activity. To this class belongs the effect that admixtures of fine alumina particles to ionic materials such as LiI, AgCl, AgBr, TlCl, CuBr increase the overall conductivity by orders of magnitude (for reviews see [30, 31, 32]). As the activation enthalpy indicates and the more detailed studies show, cations are trapped to the basic oxide's surface and mobile vacancies left [25, 30]. When the materials are interstitially conducting in the bulk as is the case for pure AgCl or AgBr, inversion layers may occur; in the case of the Schottky disordered TlCl the conductivity even changes from anion conduction in the bulk to Tl^+ conduction at the boundary. (ii) The second case (b) addresses the contact of two ionic conductors at which a redistribution over two space charge regions is thermodynamically demanded. Giant conductivity anomalies observed for compositions within the miscibility gap of AgBr:β-AgI [33] or AgCl:β-AgI [34] can be explained in this way [30]. (iii) Grain boundaries are preferential sinks for cations and anions (c). These effects can be nicely tuned by incorporating active chemicals into the boundaries. Fig. 4 shows the

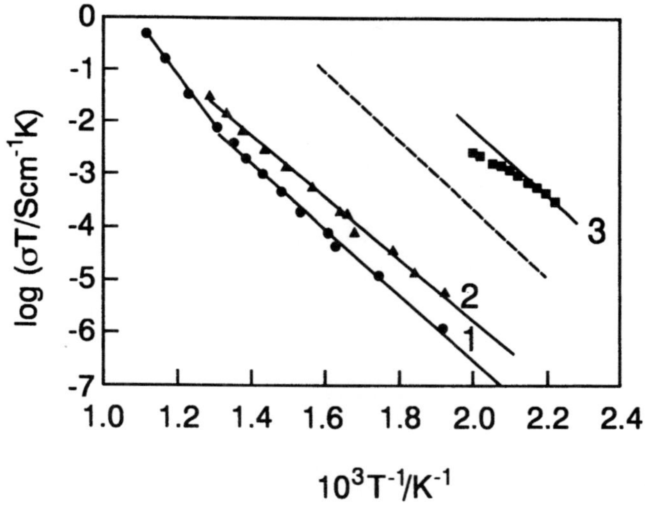

Figure 4. Ion conductivity for poly (0.2 μm) (curve 1) and nano (0.09 nm) crystalline CaF_2 (curve 3). The triangles (curve 2) give values of SbF_5 activated material [35, 36]. The dashed curve gives the space charge effect for 0.2 μm if the surface concentration is at maximum. The upper curve (curve 3) corresponds to $\sim$ 10 nm and maximum surface concentration.

effect of SbF_5 contamination on the CaF_2 grain boundary conductivity [35]. As indicated by the activation energy, F^- is trapped to this fluoride attractor leaving behind mobile vacancies which are responsible for the conductivity enhancement. (iv) Finally (Fig. 3, l. h. s., d), also fluid phases can be active: AgCl can sense ammonia gas by its effect on the surface ion conduction [37]. Like adsorbed O_2 traps electrons at the surface of semiconducting oxides, NH_3 attracts Ag^+ to the adsorption layer which leaves a vacancy behind. Let us go back to the grain boundary case. Many doped oxides exhibit depletion layers. If one measures the conductivity of a $SrTiO_3$ (Fe-doped) bicrystal perpendicular to the interface, pronounced extra resistances occur the values of which seem to increase with increasing discrepancy between interfacial core and bulk structure. Both resistance and capacitance indicate strongly decreased hole concentration in Mott-Schottky barriers (space charge potentials 200-800 mV) [38, 39, 40]. Owing to the higher charge the depletion of vacancies is more pronounced. The resulting decrease of the ionic transference number could be recently measured [41]. The simultaneous decrease of the ionic and electronic carriers which are both important for the oxygen incorporation leads to distinct chemical resistance on the oxygen diffusion which has been indeed observed in-situ by an optical technique [42]. Such Schottky barriers seem also to be of high relevance for the grain boundary resistance of Y-stabilized ZrO_2, leading to a decrease of both $V_O^{\cdot\cdot}$ and $h^{\cdot}$ [43]. Note that simultaneously the conduction electron concentration increases, which may be perceived only at extreme effects [44]. In the case of rare earth doped ceria, however, in which excess electrons play a pronounced role in the bulk (see next section), the electronic conduction is expected to be significant at such boundaries [45, 46, 28, 47].

Before we discuss nano-sized materials let us briefly see how boundary and bulk effects superimpose in polycrystalline materials: The simultaneous influences of bulk, interfacial core and space charge effects, make this superposition non-trivial even in a brick-layer-model. The situation is additionally aggravated by the fact that in both Gouy-Chapman and Mott-Schottky layers the integral of c(x) and the inverse integral of 1/c(x) are distinctly different. In other words, the electrical effect of the space charge zone itself is pronouncedly anisotropic. As shown in detail in Ref. [48], the superposition of all the effects in a bricklayer model leads to

$$\widehat{\sigma}_{\mathrm{m}} = \left[\widehat{\sigma}_{\infty}\widehat{\sigma}_{\mathrm{L}}^{\perp} + \beta_{\mathrm{L}}^{||}\varphi_{\mathrm{L}}\widehat{\sigma}^{||}\widehat{\sigma}_{\mathrm{L}}^{\perp}\right] / \left[\widehat{\sigma}_{\mathrm{L}}^{\perp} + \beta_{\mathrm{L}}^{\perp}\varphi_{\mathrm{L}}\widehat{\sigma}_{\infty}\right] \tag{11}$$

(φ_{L}: volume fraction of the interface, $\beta_{\mathrm{L}}^{\perp} = 1/3 = \beta_{\mathrm{L}}^{||}/2$ ideally). Eq. (11) describes this intercorrelation in terms of complex conductivities, but is also valid for the d.c. values.

The complex interfacial parameters $\hat{\sigma}^{||}, \hat{\sigma}_{L}^{\perp}$ are composed of core and space charge contributions [48, 30]. Eq. (11) expresses the fact that interfacial layers parallel to the current direction can bypass the bulk, both contributions being hindered by layers perpendicular to the current direction. In the impedance measurements Eq. (11) leads to two semicircles: The high frequency semicircle is approximately characterized by the bulk capacitance and the parallel combination of bulk and interfacial layer, while the low frequency semicircle is determined by capacitance (thickness) and resistance of the blocking layer to be passed. Thus, in the general case it can be that both one and the same type of interface acts short-circuiting and blocking depending on the orientation with respect to the current flow. Polycrystalline AgCl serves as a nice example: Both highly conducting space charge layers bypass the bulk, yet current constriction leads to additional blocking effects. When the ceramic is coarsened the high frequency semicircle shrinks while the low frequency semicircle is enlarged. Usually enhancement layers are perceived in the high frequency response, while depletion layers address the low frequency part. Note that for certain microstructures, however, the parallel pathways can bypass both bulk and perpendicular grain boundary pathways. In the case that the conduction type changes from electronic to ionic or vice versa when we compare bulk with grain boundary, stoichiometric polarization effects appear which can become very subtle [49].

4. NANOIONICS: NANO-SIZED CRYSTALS AND FILMS

4.1 General Considerations

Research on the equilibrium structures of $(NaCl)_n$ oligomers show [50] that already for $n > 10$ the rock-salt structure is established with bulk and surface properties not too far from values of the massive crystals. Below that limit (corresponding to sizes of the order of 1 nm) the regime of cluster chemistry is established with distinctly different structures and properties.

Even though the massive structure is readily achieved, the fraction of boundaries and thus their impact is fairly high for structures even on the order of 10 nm and more.

Fig. 5 displays (according to Ref. [1]) the fraction of corner, edge, plane and bulk atoms as a function of the total number of atoms for cubic crystals: While $N_V = (N^{1/3} - 2)^3$ is the number of particles sitting in the interior of the cube, the number of particles sitting in planes (but not in edges or corners) is $N_P = 6(N^{1/3} - 2)^2$, the number of particles sitting in edges (but not in corners) is $N_E = 12(N^{1/3} - 2)$

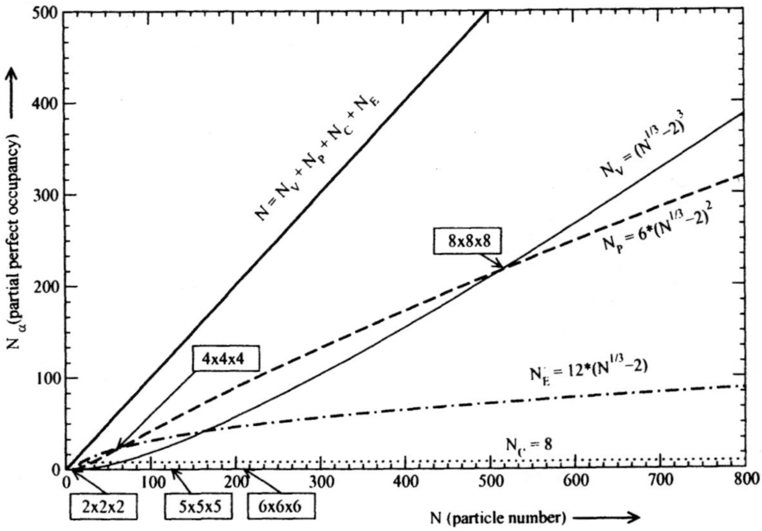

Figure 5. The number of particles (i) sitting only in the interior (N_V), (ii) sitting in the planes but not in edges or corners (N_P), (iii) sitting in edges but not in corners, (iv) sitting in corners of cubic crystals ($N^{1/3} \times N^{1/3} \times N^{1/3}$) as a function of the particle number N [1].

and the number of particles in corners is equal to $N_C = 8$. In Fig. 5 N_C, N_E, N_P, N_V and their sum N are plotted as a function of the cube size. The $2 \times 2 \times 2$ cube has only corners; for the $4 \times 4 \times 4$ cube it holds that $N_E = N_P > N_V$, only for cubic clusters larger than the $8 \times 8 \times 8$ cube $N_V > N_P$ and still for a $10 \times 10 \times 10$ cube, i. e. for 1000 particles, approximately half of the material does not sit in the interior. Thus "trivial effects", i. e. effects which just stem from the geometrically changed fraction of particles sitting in different energetic positions, can become very pronounced even if we assume the local specific free energies of the massive crystal. As an orientation aid a cube of thickness of 300 Å may be considered which corresponds, for a mean particle separation of 5Å, to a $60 \times 60 \times 60$ crystal; there the surface volume fraction V^s/V is still 10%. Similar considerations hold for a general Wulff crystal (see below) [1]. A striking example is the enormous depression of the melting point of gold particles (it changes by several hundreds of degrees if the mean radius $\bar{r}$ varies between 200 Å and 20 Å) which has been fully explained by using parameters of massive gold [51, 52].

Another example is the e. m. f. measured for the cell [53]

$$\mathrm{Cu\,(macro)\,|CuSO_4\,(H_2SO_4)|\,Cu\,(nano)}$$

which reflects the different surface contributions on both sides (see also Ref. [54] for Ag(nano) deposition). For these considerations we have to assume nano-crystals of equilibrium shape, for which the chemical potential of the substance μ_{MX} (here MX=Cu) is given by

$$\mu_{MX} = \mu_{MX}^{\infty} + \frac{2\bar{\gamma}}{\bar{r}} V_m \tag{12}$$

(μ_{MX}^{∞}: chemical potential of MX for infinitely large crystals, $\bar{\gamma}$ and $\bar{r}$ refer to the tension (γ_j) and the distance from the center (h_j) of the individual crystallographic planes (j) and obtained from these by area-averages (a_j). For Wulff crystals the ratio γ_j/h_j is a constant. V_m is the molar volume.) The fraction (by volume) of the surface shell characterized by a thickness $|s|$ related to the total volume of a Wulff crystal follows as $3|s|/\bar{r}$ (cubic crystal $(6N^{2/3} - 12N^{1/3} + 8)/N \simeq 6N^{2/3}/N = 6|s|/L$). The significance of Wulff shape considerations is quite restricted. Large crystals usually do not reach the equilibrium form for kinetic reasons, while the equilibrium form of small crystals deviates from the Wulff shape according to the increasing influence of edges. The condition $\gamma_j/h_j = \text{const}$ for each surface plane j has then to be replaced by $(\gamma_j + \Sigma_k \chi_k \frac{\partial L_k}{\partial a_j})/h_j = \text{const}$ (χ_k, L_k: edge tension and edge length other than the k'th edge) [55]. Additionally the influence of the substrate must usually not be ignored [56].

In a certain sense Eq. (12) offers an example for the distinction between trivial and proper size effects; here as far as energetics are concerned. As long as μ_{MX} changes with size while $\bar{r}$ and $\bar{\gamma}$ stay constant we meet a trivial (here energetic) size effect, while a proper size effect comes into play when $\bar{\gamma}$ changes with $\bar{r}$ (see e.g. cluster chemical regime).

Similarly, we will distinguish between trivial and proper size effects in the context of defect chemistry. Owing to the increasing fraction of the interfacial zone, interfacial core and space charge effects become increasingly important when the size is reduced. Proper size effects emerge when the crystal becomes comparable with characteristic boundary length scales.

4.2 Trivial and Proper Size Effects on Defect Chemistry

Trivial size effects leave the local defect chemistry unchanged and enter the effective conductivity via a changed φ_L value only (see Eq. (11)). Let us for simplicity consider a film evaporated on a substrate the parallel conductance of which we measure as a function of the thickness L (we will assume that the interface to the substrate exhibits an enhanced conductivity similar to the solid/gas interface, see Fig. 6). If the film thickness is so large that its influence is negligible, $Y^{||}$ is linear in L with

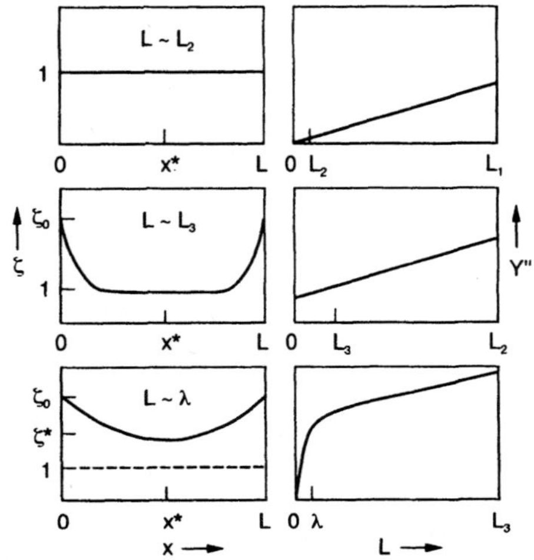

Figure 6. Conduction profiles and parallel conductance for various thicknesses of thin films [57]. (L_1, L_2, L_3 are arbitrary thicknesses characterizing the length scales ($L_3 \ll L_2 \ll L_1$). ζ: concentration divided by bulk concentration)[30].

a zero-intercept. Owing to

$$Y^{||} = |z_1|Fu_1 \left[2\sqrt{2\varepsilon RTc_{10}} + c_\infty L\right], \tag{13}$$

a non-zero intercept becomes visible for small thicknesses and increasingly important. At $L \simeq 4\lambda$ the space charge influence prevails completely. So far the mean space charge conductivity $|z_1|Fu_1 \sqrt{2\varepsilon RTc_{10}}/2\lambda \propto \sqrt{c_{10}c_{1\infty}}$ remained unaffected (see below Fig. 7b r. h. s.). A local size effect occurs if the spacing of the interfaces is additionally re-

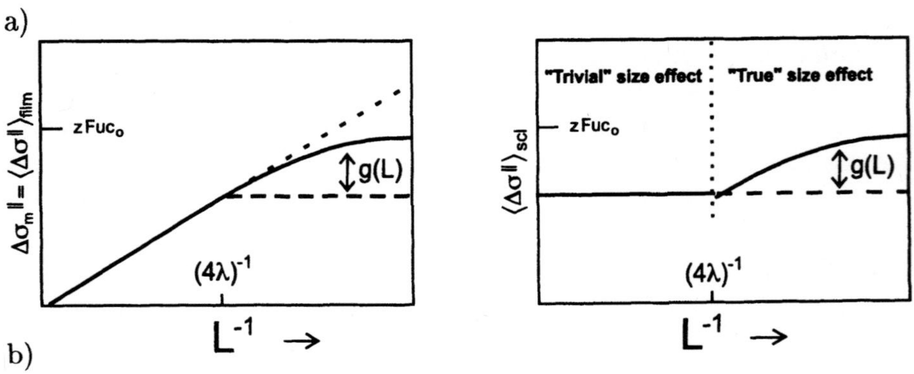

Figure 7. a) Area related space charge conductance per film thickness representing the measured excess space charge conductivity, as a function of inverse spacing. b) Area related space charge conductance per boundary thickness representing the mean specific excess space charge conductivity, as a function of inverse spacing [4].

duced. $\sqrt{c_{10}}$ in the above formula has to be replaced by $\sqrt{c_{10} - c_1^*(L_1, c_{10})}$ where c^* measures the concentration in the film center and is correlated with L and c_{10} via

$$2\sqrt{\frac{c_{1\infty}}{c_1^*}}\left(\mathcal{E}\left(\frac{c_{1\infty}}{c_1^*};\frac{\pi}{2}\right)-\mathcal{E}\left(\frac{c_{1\infty}}{c_1^*},\arcsin\sqrt{c_1^*/c_{10}}\right)\right)=\frac{L}{2\lambda}. \tag{14}$$

$\mathcal{E}$ represents elliptical integrals of the first kind [57]. The influence of c^* in shown in Fig. 8. The impact on $Y^{||}$ becomes now smaller than extrapolated since the total surface charge decreases. The profiles inside become flatter. Eq. (14) predicts that the value c_{10} is eventually reached throughout the sample: If $L/2\lambda$ approaches zero in Eq. (14),

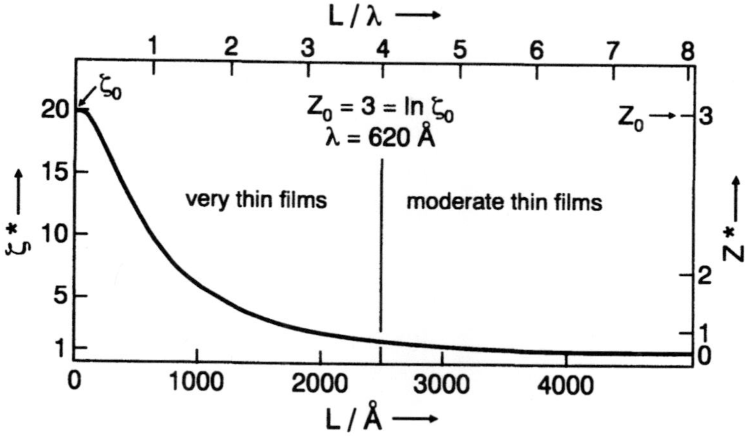

Figure 8. The concentration (conductivity) enhancement in the sample center as a function of thickness ($\zeta_0 = 20, \lambda = 620$ Å) ($z \equiv \ln\zeta, \zeta \equiv c/c_\infty$) [57].

$\arcsin\sqrt{c_1^*/c_{10}}$ approaches $\pi/2$ and hence $c_1^* \longrightarrow c_{10}$. The effects of trivial and true size effect are best illustrated in Figs. 7a and 7b. Fig. 7a displays the measured conductivity obtained by relating the conductance ($Y^{||}$) to the thickness, plotted as a function of $1/L$. If the bulk contribution can be ignored, it is simply:

$$\sigma_m^{||} \simeq \frac{2|z_1|Fu_1\sqrt{2RTc_{10}}}{L}, \tag{15}$$

and continuously increases with decreasing L since the fraction of insulating bulk diminishes. At $L \simeq 4\lambda$ the film effectively represents a single space charge zone. Nonetheless, further shrinkage of L still increases $\sigma_m^{||}$ since the local defect density in the film grows towards the limit $|z_1|Fu_1c_{10}$ (see Fig. 8). Fig. 7b displays the development of the local mean space charge conductivity, i. e. $Y^{||}/2\lambda$. This quantity is enhanced compared to the local mean conductivity at $L \simeq 4\lambda$ by

$$\text{nano-size factor} \equiv g(L \leqslant 4\lambda) = \frac{4\lambda}{L}\sqrt{\frac{c_{10} - c_1^*}{c_{10}}}, \tag{16}$$

for $L \simeq 4\lambda$ the nano-size factor is unity, and adopts increasingly higher values for $L < 4\lambda$. The change at $L < 4\lambda$ indicates a non-trivial size effect.

Further size effects are to be expected if structural changes occur. This is, at the latest, the case if L is of the order of the shell thickness s. (Indeed the decrease below typical thicknesses of the order of 1 nm has been found to lead to distinct anomalies in the properties of nano-crystals of some noble metals [58].) Elastic effects of longer range let emerge such size effects at larger spacings. All these phenomena affect the levels in the "energy level diagrams". Note that de-localized electrons perceive the interfaces much earlier since being effectively more extended than ions. In pronounced semiconductors these quantum mechanical confinement effects usually prevail over electronic space charge effects.

In addition, electrons and holes can not be sufficiently separated in nano-crystals in order to allow the neglect of interactions. Thus both counter-acting effects are generally important in semiconductors. Hence the band-gap E_g has been predicted [59] to depend on the size (sphere radius r) as

$$E_g = E_{g,\infty} + \frac{\hbar^2\pi^2}{2r^2}\left[\frac{1}{m_n} + \frac{1}{m_p}\right] - \frac{1.8e^2}{\varepsilon_2 r} + \langle\frac{e^2}{r}\sum_{n=1}^{\infty}\alpha_n\left(\frac{\nu}{r}\right)^{2n}\rangle \tag{17}$$

(E_g: band gap; $E_{g,\infty}$ is band gap for $r \longrightarrow \infty$; m_n, m_p are the effective masses of e', $h^{\cdot}$, ε_2 is the dielectric constant of the crystal. The last term arises from dielectric screening and refers to the surrounding, characterized by α and n [59].). Most of the points discussed so far have been highlighted for one-dimensional systems. For three-dimensional systems intriguing effects occur that involve the particular roles of edges and corners. In nanocrystalline systems the size dependence of thermodynamic properties makes differently large particles thermodynamically unlike, which leads to peculiar conductivity effects, the consideration of which is beyond the scope of this paper [60].

Phenomena of particular significance are boundary phase transformations. Owing to the special energetic and compositional situation, phase transformation can occur at temperatures different from the bulk (see also modified defect chemistry), or even transformations which are not found in bulk structures at all may become apparent [61]. Phase transformations which are influenced by the distance of the neighboring interfaces are particularly interesting. (Since the structures met in cluster chemistry are not identical to the structure of the surface layer in crystals, the above-mentioned transition from NaCl rock salt structures to special cluster structures when n in $(NaCl)_n$ becomes smaller than 10 represents such a type of phase transformation).

5. NANO-SIZED ION AND MIXED CONDUCTORS: EXAMPLES

5.1 Nano-Sized Li Conductors

Even though numerous strategies are known that lead to nano-crystals, it is much more difficult to produce compacted nanocrystalline samples the conductance of which can be reliably studied. (Useful methods are hot-pressing and explosive compaction. In the latter way an all solid state nano-crystalline Li-battery could be fabricated [62].) From that point of view films are advantageous. LiI films on SiO_2 substrate prepared in Ref. [63] are not really nano-sized and did not exhibit space charge anomalies, yet they were interesting from a structural viewpoint: Thin films grow hexagonal even though massive LiI is cubic (see Fig. 9, top). Ion conductor films which are much thinner have been grown by

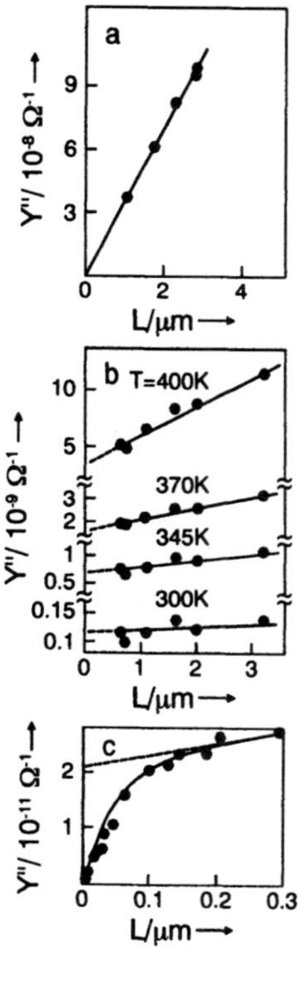

Figure 9. Examples of the thickness dependence of the parallel conductance corresponding to Fig. 6: Top: LiI on SiO_2, center: AgCl on mica, bottom: LiI on Al_2O_3 [30].

Schreck et al. [64] on a sapphire substrate (bottom picture of Fig. 9). Even though the $Y^{||}$ characteristic is of the type of the bottom picture in Fig. 6 and even though it can be well fitted by finite space charges [57], the parameters obtained suggest that the results are obscured by islands formation and percolation problems [57, 30]. Nanocrystalline Li-conductors in the form of monophasic or biphasic materials have been studied in Refs. [65, 66]. Even though significant effects on the transport is reported, no size anomalies seem to occur. Studies on Li_xTiS_2 [65] give evidence on interfacial conductivity which is distinctly different from the behavior of the amorphous material; studies on $Li_2O : B_2O_3$ essentially stress the effect of size-reduction on the percolation behavior [66].

Since the major interest in Li materials is devoted to Li batteries, it should be emphasized that both capacity and speed of storage of Li will sensitively depend on the crystal size. Not only is the diffusion length drastically reduced, also the diffusion constants are expected to change; in addition at $L \leq 4\lambda$ the differences between electrical capacitor, supercapacitor and chemical capacitor are getting blurred [7]. Two recent findings are interesting in this context. Tarascon [67] showed that CoO can be used in a reactive way for anodes: (i) The Li_2O crystals formed establish a thickness of about 1 nm and can be reversibly reduced. (ii) Electronically insulating lithium iron phosphate [68] can be used as Li cathode material, if the material is nano-crystalline and in contact with carbon; here the grain-size is also about 1 nm [69].

5.2 Nanocrystalline CeO_2

The most intensively studied nanocrystalline ion conductor is CeO_2. If appropriately doped, usually with rare earth elements, it exhibits a high oxygen vacancy conductivity and distinctly lower electronic conductivity [70]. If the material is pure, the conductivity is predominantly electronic ($[e'] \simeq 2\,[V_O^{\cdot\cdot}]$).

The experimental facts are the following: Nanocrystalline ceria shows an increased conductivity compared to coarsened material. Owing to the distinct decrease with P_{O_2} this has been attributed to n-type electronic conduction (with typical dependencies $\partial \ln \sigma / \partial \ln P$ of -1/6 or (-1/4)) [71, 72, 73].

Two extreme models are imaginable and both have been proposed:

a) The first model [71] attributes the effect solely to the disorder in the interfacial core. In order for it to be consistent one has to assume such a high disorder that space charge effects are negligible. Owing to reduced formation energies of the defects there ($V_O^{\cdot\cdot}, e'$) [74], i.e. owing to a strongly modified mass action constant of oxygen incorporation, the concentration of both ionic and electronic carriers is considered to be enhanced. If we assume electroneutrality in the core layer, i.e. space charge negligible to lead intrinsic disorder, a P_{O_2}-dependence of -1/6 follows. The observed depression of the ion conductivity then must be ascribed to a lower mobility in the core and a reduced concentration in the bulk. The latter might be due to the following size effect: With increasing interfacial density so much segregation can occur that the grain interior becomes progressively cleaner. As a result the impurity concentration in the bulk (and hence also the Debye length) will change with grain size. b) The second model concentrates on the space charge effect and assumes negatively charged space charge zones, i.e. a positive excess charge in the interfacial core [45]. A recent study of polycrystalline CeO_2 essentially confirms this sign of the space charge layers necessarily leading to an increased electronic conductivity combined with the depletion of vacancies [41]. Thus σ_{eon} should rise and σ_{ion} should fall. Very recent careful experiments by Kim et al. [28] allowed the separation of ionic and electronic contributions as well as bulk and space charge effects. These investigations clearly favor the space charge model. The detailed analysis is quite involved and given in Ref. [28].

Spectroscopic studies on nanocrystalline CeO_2 [73] address the above discussed points of confinement and excitation effects on the electronic carriers. Blue and red shifts occurring in CeO_2 have been explained by confinement and Coulomb effect according to Eq. (10). Note that edge and corner effects are neglected in the analysis, which are, however, expected to play a significant role [75].

5.3 Nanocrystalline CaF_2

Nanocrystalline CaF_2 seems to be much better understood. Both order of magnitude and activation energy strongly suggest space charge effects characterized by increased vacancy concentrations in the space charge zones [36] (see Fig. 4). The increased conductivity is derived from the high frequency semicircle. When coarsening the sample this semicircle is enlarged. Opposite to the AgCl example discussed above, the low frequency semicircle increases, too. This finds its explanation in the TEM pictures revealing that large pores open up which obviously lead to increased constriction effects. The explanation of the σ-values of nano-CaF_2 succeeds without the need to bring finite space charges into the play. This is also consistent with the impurity content which points towards a very low Debye length ($\sim 1\,\mathrm{nm} \ll 9\,\mathrm{nm} \sim$ grain size). However, the surface effect must be assumed to be almost in the saturation limit, i.e. much higher than for polycrystalline material. It is, however, expected that the surface effect strongly depends on the "degree of influence" and thus on grain size, preparation etc.

5.4 Nano-Sized CaF_2-BaF_2 Heterolayers

Recently CaF_2-BaF_2 heterolayers have been prepared by molecular beam epitaxy [76, 77]. The spacing has been tuned from $1\mu m$ to values certainly below the Debye-length (1 nm). Each heterojunction is reflected in an increase of conductivity. For moderate spacings ($\gtrsim 4\lambda$) the conductivity can be quantitatively explained by space charge effects. At very low spacings the conductivity does not saturate at $L \simeq 4\lambda$ (see Fig. 10), i.e. the local boundary conductivity increases indicating finite boundary conditions. Moreover, the conductance increases even more steeply with increasing interfacial density. The detailed explanation of this must go beyond the treatment above and include elastic and defect interactions. Although the conductivity phenomena can be best explained by a partial transfer of F^- from BaF_2 to CaF_2, a segregation of F^- into the proper interface should not be excluded since it leads to a lowering of the local misfit. Whatever the detailed picture is, it is clear

Figure 10. Comparison of the conductivity profiles in the semi-infinite space charge and mesoscale situations. The concentration or (parallel) conductivity profiles are sketched for the semi-infinite space charge situation (period $> 8\lambda$, left), and for the mesoscale situation (period $< 8\lambda$, right) in which the space charge regions overlap and bulk values are exceeded even in the centers of the individual layers [76].

that in this way artificial ion conductors can be generated, heterostructures in which the individual layers have lost their individuality and a novel system property emerged.

5.5 AgI-Al_2O_3 Composites: Interfacial Phase Transformation and High Conductivities in Stacking Faults Arrangements

A qualitatively similar explanation has been recently brought into play to explain the anomalous Ag^+ conductivities in composites of β-AgI and Al_2O_3 [78]. Whilst the conductivity effect in heterogeneously doped AgCl or AgBr [30] can be satisfactorily interpreted by semi-infinite space charges, the conductivity effects in β-AgI:Al_2O_3 [79] two phase mixtures are far too high. Yet, trends and dependencies are similar and suggest related phenomena. The spacing of alumina profiles is not small enough to generate finite space charges. A closer inspection of the X-ray results shows a peculiar situation. The formation of stacking faults [78] at the interface is indicated (Fig. 11). Interestingly, the stacking faults can

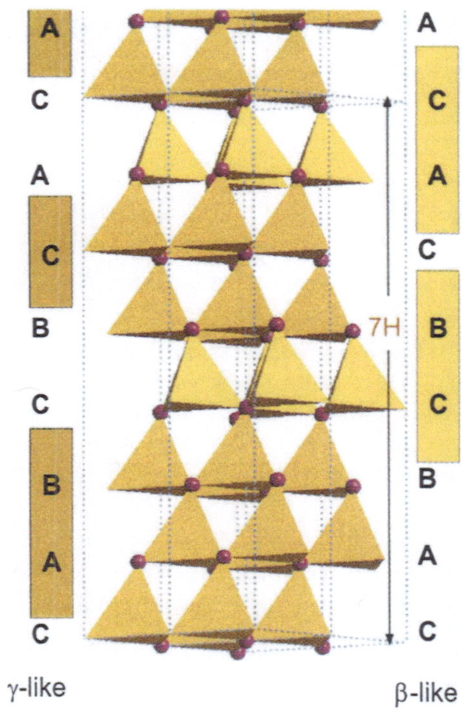

Figure 11. 7H structure represented by close packed layers of tetrahedral sites occupied by silver ions. Partial γ-AgI like (left) and β-AgI like (right) stacking sequences are indicated in colored regimes [78, 6].

be conceived as heterolayers of γ- and β-AgI on an atomic level. Since (metastable) two phase mixtures of β- and γ-AgI exhibit conductivity anomalies, a charge redistribution similar to the previous example may occur.

6. PERSPECTIVES OF NANO-IONICS AND SOFT MATERIALS SCIENCE

These few examples clearly revealed the importance of the exploration of the nano-scale in view of its potential to modify ion conductivity in an overall sense, even locally. It is clear that these interfacially dominated materials are morphologically metastable, sometimes, however, metastable enough to be stationary even at elevated temperatures (see above examples). Nano-structuring which introduces multi-functionality, can thus be of relevance also for comparatively high

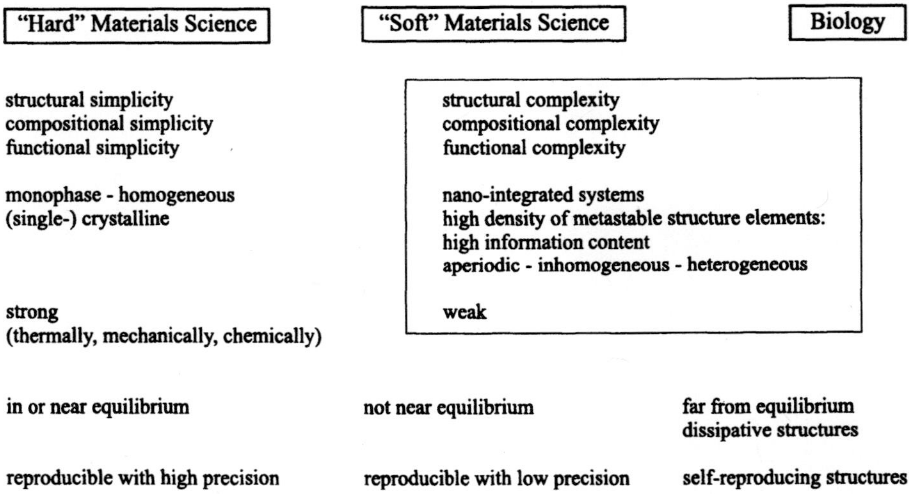

Figure 12. The structuring on the nano-scale corresponds to a high density of metastable structure elements characterizing a "soft materials science" with specific advantages and disadvantages: the figure gives a demarcation of "soft materials science" with respect to "hard" materials science and the other extreme case of biology, and stresses the common points with the latter (see box) [6].

temperatures. Yet, the interfacial effects discussed above are particularly significant at lower temperatures. Since it is desired to extend the operational range of ionic devices down toward room temperature, the significance of the exploration of the structurability on the nano-scale in which chemistry and physics meet in an exciting way, is highly timely. Structuring the nano-scale is one of the key successes of biology. Apart from the use of dissipative structures or even self-replicating structures which is intimately connected with specific biological chemistry, materials science even in the purely inorganic world can learn a lot from biology in terms of compositionally and functionally structuring the nano-world (see Fig. 12). This structuring is equivalent to the introduction of metastable structure elements and corresponds to the

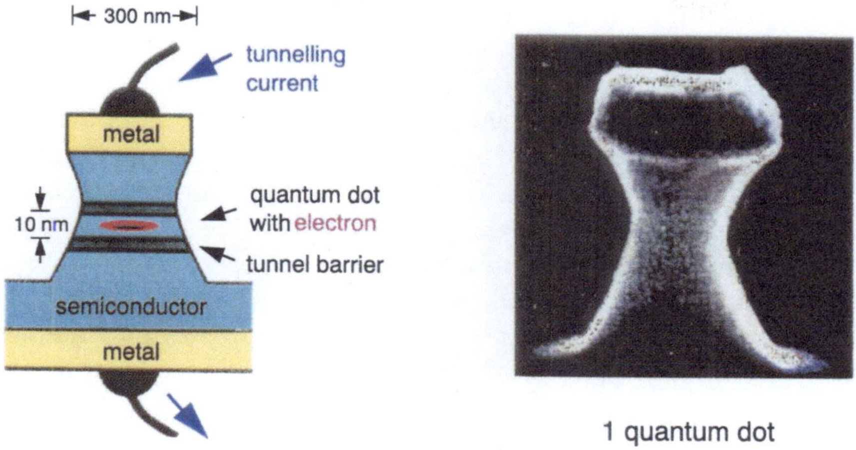

Figure 13. Quantum dots as artificial single electron devices [80, 6].

implementation of a high information content which is not possible in the thermodynamic equilibrium. Figs. 13, 14 show two examples: one (Fig. 13) is the detailed structuring of quantum dots which represent a true nano-scale device. The other (Fig. 14) is the situation occurring in

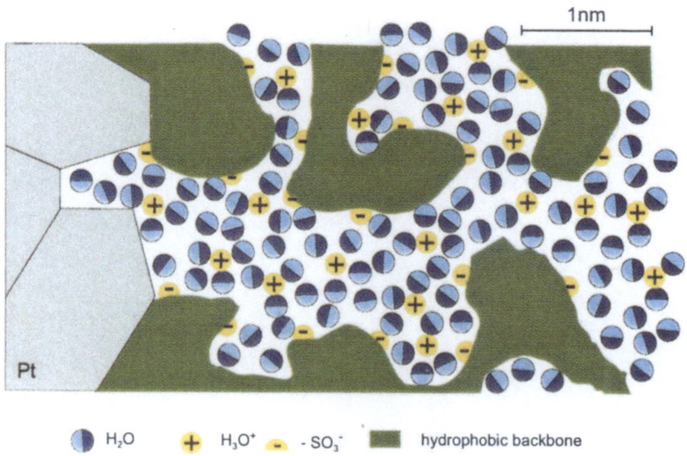

Figure 14. Mesoscopic arrangement of acidified water channels in the PEEK-polymers [81, 6].

a PEM fuel cell at the electrode: The ion exchange membrane is composed of an organic backbone and water filled channels. These sulfonic acid surface groups sitting on the backbone dissociate and the protons can be transferred away. The water channels are so tiny that electrical fields are almost negligible (finite space charges!). The Pt-particles act as current collectors and catalysts, almost comparable to organelles in biological cells. In future — and this is triggered by the possibility to manipulate and to analyze the nano-scale to an advanced degree — such a "soft materials science" will certainly become very important. Of course leaving the safe ground of hard materials science (master example: Si-technology) has its trade-offs in giving up the extreme stabilities and reproducibilities which can be achieved there. There are many properties for which high thermal or chemical stability or the enormously accurate reproducibility of structure and property of non-complex systems will remain indispensable such that both poles — hard and soft materials science — will characterize future developments.

References

[1] J. Maier, Electrochemistry **68** (6) (2000) 395.

[2] J. Schoonman, Solid State Ionics, Proc. SSI 2001 (S. Badwal (ed.)), to be published.

[3] J. Schoonman, Solid State Ionics **135-137** (2000) 5.

[4] J. Schoonman, Solid State Ionics, Proc. ISSFIT 6 (J. Molenda et al. (eds.)), to be published.

[5] J. Maier, Solid State Ionics **135** (1-4) 2000 575.

[6] J. Maier, Solid State Ionics, Proc. SSI 2001 (S. Badwal (ed.)), to be published.

[7] J. Maier, Solid State Ionics, Proc. ICMAT-2001 (B. V. R. Chowdari (ed.)), to be published.

[8] J. Maier, Solid State Ionics, Proc. ISSFIT 6, to be published.

[9] Special Issue on Nanostructured Materials for Energy Applications, Y.-M. Chiang, (ed.), J. Electroceramics **3** (1997) pp. 205.

[10] H. L. Tuller, Solid State Ionics **131** (2000) 143.

[11] P. Moriarty, Rep. Progr. Phys. **64** (2001) 297.

[12] J. Maier, Angew. Chem. Int. Ed. Engl. **32** (3) (1993) 313; Angew. Chem. Int. Ed. Engl. **32** (4) (1993) 528.

[13] J. Maier, Chemistry — A European Journal, in press.

[14] J. Maier, Solid State Ionics **143** (2001) 17.

[15] J. Maier, Solid State Ionics **131** (1, 2) (2000) 13.

[16] J. Maier, *Festkörper — Fehler und Funktion: Prinzipien der Physikalischen Festkörperchemie*, B. G. Teubner Verlag, Stuttgart, 2000.

[17] S. Stotz and C. Wagner, Ber. Bunsenges. Phys. Chem. **70** (1966) 781.

[18] H. Iwahara, T. Esaka, H. Uchida, N. Maeda, Solid State Ionics **314** (1981) 259.

[19] K. Sasaki and J. Maier, J. Appl. Phys. **86** (10) (1999) 5422; J. Appl. Phys. **86** (10) (1999) 5434.

[20] A. B. Lidiard, Phil. Mag. **1** (1956) 1.

[21] N. Hainovsky and J. Maier, Phys. Rev. B **51** (22) (1995) 15789.

[22] B. A. Huberman, Phys. Rev. Letters **32** (1974) 1000.

[23] J. Maier and W. Münch, Z. Anorg. Allg. Chem. **626** (2000) 264.

[24] J. Jamnik, J. Maier, S. Pejovnik, Solid State Ionics **75** (1995) 51.

[25] J. Maier, J. Electrochem. Soc. **134** (1987) 1524.

[26] G. Gouy, J. Physique **9** (1910) 457; D. L. Chapman, Phil. Mag. **25** (1913) 475.

[27] S. M. Sze, *Semiconductor Devices*, John Wiley & Sons, New York, 1985.

[28] S. Kim, J. Fleig and J. Maier, Proc. 4th Int. Symp. Ionic and Mixed Conducting Ceramics, in press; S. Kim, J. Maier, in preparation.

[29] J. Maier, Solid State Phenomena **67-68** (1999) 45.

[30] J. Maier, Prog. Solid St. Chem. **Vol. 23** (3) (1995) 171.

[31] N. J. Dudney, Ann. Rev. Mat. Sci. **19** (1989) 103.

[32] P. Knauth, J. Electroceramics **5** (2) (2000) 111.

[33] K. Shahi, J. B. Wagner, Appl. Phys. Lett. **37** (1980) 757.

[34] U. Lauer and J. Maier, Ber. Bunsenges. Phys. Chem. **96** (1992) 111.

[35] Y. Saito and J. Maier, J. Electrochem. Soc. **142** (9) (1995) 3078.

[36] W. Puin, S. Rodewald, R. Ramlau, P. Heitjans, J. Maier, Solid State Ionics **131** (1, 2) (2000) 159.

[37] U. Lauer, J. Maier, W. Göpel, Sensors and Actuators B, **2** (1990) 125.

[38] R. Waser, J. Am. Ceram. Soc. **74** (1991) 1934.

[39] Y.-M. Chiang and T. Tagaki, J. Am. Ceram. Soc. **73** (1990) 3278; Y.-M. Chiang, E. B. Lavik, I. Kosacki, H. L. Tuller, J. Y. Ying, J. Electroceramics **1** (1997) 7.

[40] I. Denk, J. Claus and J. Maier, J. Electrochem. Soc. **144** (10) (1997) 3526.

[41] X. Guo, J. Fleig and J. Maier, J. Electrochem. Soc. **148** (9) (2001) J50.

[42] M. Leonhardt, J. Jamnik and J. Maier, Electrochemical and Solid-State Letters **2** (7) (1999) 333.

[43] X. Guo and J. Maier, J. Electrochem. Soc. **148** (3) (2001) E121.

[44] R. Waser and R. Hagenbeck, Acta Mater. **48** (2000) 797.

[45] A. Tschöpe, Solid State Ionics **139** (2001) 267.

[46] X. Guo and J. Maier, J. Electrochem. Soc., in preparation.

[47] S. Kim and J. Maier, in preparation.

[48] J. Maier, Ber. Bunsenges. Phys. Chem. **90** (1986) 26.

[49] J. Jamnik, Solid State Ionics, Proc. ISSFIT 6, to be published; X. Guo and J. Maier, in preparation.

[50] T. P. Martin, *Festkörperprobleme (Advances in Solid State Physics)*, **Vol. XXIV**, Vieweg, Braunschweig, 1984, 1; Phys. Reps. **95** (1983) 167.

[51] Ph. Buffat, J.-P. Borel, Phys. Rev. A **13** (1976) 2287.

[52] C. J. Coombes, J. Phys. F. **2** (1972) 441.

[53] Ch. P. Gräf, U. Heim, G. Schwitzgebel, Solid State Ionics **131** (2000) 165.

[54] S. Villain, P. Knauth, G. Schwitzgebel, J. Phys. Chem. B **101** (1997) 7452.

[55] A. I. Rusanov, *Phasengleichgewichte und Grenzflächenerscheinungen*, Akademie-Verlag, Berlin, 1978.

[56] R. Kaischew, Bull. Acad. Bulg. Sci. Phys. **2** (1951) 191.

[57] J. Maier, Solid State Ionics **23** (1987) 59.

[58] See references in [1].

[59] L. E. Brus, J. Phys. Chem. **90** (1986) 2558.

[60] I. Lubomirsky, J. Fleig, J. Maier, manuscript.

[61] R. Lipowsky, *Phasenübergänge an Oberflächen* (IFF–Ferienkurs), S. 9.1, Forschungszentrum Jülich GmbH, 1993; Springer Tracts in Mod. Phys., **Vol. 127**, in preparation; H. Dosch, *Critical Phenomena at Surfaces and Interfaces: Evanescent X-Ray and Neutron Scattering*, Springer Tracts in Mod. Phys., **Vol. 126**, Springer, 1992.

[62] J. Schoonman, Solid State Ionics, Proc. ISSFIT 6, to be published.

[63] B. Wassermann, T. P. Martin, J. Maier, Solid State Ionics **28-30** (1988) 1514.

[64] E. Schreck, K. Länger, K. Dransfeld, Z. Physik B **62** (1986) 33.

[65] R. Winter, P. Heitjans, J. Phys. Chem. B (2001) in press.

[66] P. Heitjans, H. E. Roman, A. Bunde, Phys. Rev. Lett. **84** (2001) 2889-2892.

[67] J. M. Tarascon, Solid State Ionics, Proc. SSI 2001, to be published.

[68] A. K. Padhi, K. S. Nanjundaswamy, and J. B. Goodenough, J. Electrochem. Soc. **144** (1997) 1188.

[69] N. Ravet, J. B. Goodenough, S. Besner, M. Simoneau, P. Hovignon, and M. Armand, Abstract 127, The 1999 Joint International Meetings Abstracts, Honolulu, HI, Oct. 17-22, 1999; M. Armand, Solid State Ionics, Proc. SSI 2001, to be published.

[70] H. L. Tuller, A. S. Nowick, J. Phys. Chem. Solids **38** (1977) 859.

[71] Y.-M. Chiang, E. B. Lavik, I. Kosacki, H. L. Tuller, J. Y. Ying, Appl. Phys. Lett. **69** (2) (1996) 185.

[72] I. Kosacki and H. U. Anderson, Ionics **6** (2000) 294; J. H. Hwang, T. O. Mason, Z. Phys. Chem. **207** (1998) 21.

[73] I. Kosacki, T. Suzuki, H. U. Anderson, in: J. P. Singh, N. P. Bansal and K. Niihara (eds.), Innovative Processing and Synthesis of Ceramics, Glasses and Composites III, Ceramic Trans. 108 (2000) 275; I. Kosacki, V. Petrovski, H. U. Anderson, Appl. Phys. Lett. **74** (1999) 341.

[74] T. X. T. Sayle, S. C. Parker, C. R. A. Catlow, Surf. Sci. **316** (1984) 329.

[75] P. V. Sushko, A. L. Shluger, C. R. A. Catlow, Surface Science, **450** (2000) 153.

[76] N. Sata, K. Eberman, K. Eberl and J. Maier, Nature **408** (2000) 946.

[77] N. Sata, N. Y. Jin-Phillipp, K. Eberl and J. Maier, Solid State Ionics, Proc. SSI 2001, to be published.

[78] J.-S. Lee, S. Adams, J. Maier, J. Electrochem. Soc. **147** (6) (2000) 2407.

[79] K. Shahi and J. B. Wagner, Jr., J. Electrochem. Soc. **128** (1981) 6.

[80] K. von Klitzing, *Max-Planck-Jahrbuch 1996*, Verlag Vandenhoeck & Ruprecht, Göttingen, 1996, pp. 111.

[81] K. D. Kreuer, in: B. V. R. Chowdari, K. Lal, S. A. Agnihotry, N. Khare, S. S. Sekhon, P. C. Srivastava S. Chandra (Eds.), Solid State Ionics: Science & Technology, World Scientific Publishing Co., Singapore, 1998, pp. 263.

IMPEDANCE/DIELECTRIC SPECTROSCOPY OF NANOCERAMICS

T. O. Mason, J. –H. Hwang*, N. Mansourian-Hadavi, G. B. Gonzalez, B. J. Ingram, and Z. J. Homrighaus

Northwestern University, Department of Materials Science and Engineering, 2225 North Campus Drive, Evanston, IL 60208, USA

ABSTRACT

With proper attention to experimental design (i.e., electroding, cabling, stray apparatus immittances, etc.) impedance/dielectric spectroscopy is a powerful tool to study the electrical properties of nanoscale electroceramics. This study focuses on bulk non-ferroelectric materials (ZnO, CeO_2, TiO_2) and their frequency-dependent AC electrical properties, taken from a variety of literature sources. In particular, it is shown how to separate effective grain boundary and grain interior resistivities and also the effective capacitances associated with each region in the microstructure. This is possible even when Nyquist plots ($-Z_{im}$ vs. Z_{re}) without frequency markers are the only data supplied. A modified brick layer model (BLM) can be used to analyze the impedance/dielectric properties of nanoscale ceramics.

GLOSSARY OF TERMS

A .... sample cross sectional area
BLM .... Brick Layer Model
C_{gb} .... grain boundary capacitance
C_{gc} .... grain core capacitance
C_p .... parallel capacitance of apparatus
C_{re} .... real component of capacitance
$C_{re}(hi\nu)$ .... real component of capacitance at high frequency
$C_{re}(lo\nu)$ .... real component of capacitance at low frequency
C_{sgb} .... specific grain boundary capacitance
D .... grain core size
D' .... grain size
δ .... grain boundary width
DIC .... Dispersed Ionic Conductor
ε_{gb} .... grain boundary dielectric constant
ε_{gc} .... grain core dielectric constant
I/DS .... Impedance/Dielectric Spectroscopy
L .... inductance of leads
l .... sample interelectrode spacing
n_{gb} .... grain boundary density

ν_{top} .... frequency at the top of the arc
R_{agb} .... areal grain boundary resistance
R_c .... contact resistance
R_{gb} .... grain boundary resistance
ρ_{gb} .... effective grain boundary resistivity
R_{gc} .... grain core resistance
R_s .... series resistance of leads
ρ_{sgb} .... specific grain boundary resistivity
ω .... angular frequency
ω_{top} .... angular frequency at the top of the arc
Y .... admittance
Y_o .... admittance pre-factor
Z .... impedance
Z_{im} .... imaginary component of impedance
Z_{re} .... real component of impedance
Z_{total} .... combined impedance of grain core and boundary contributions

INTRODUCTION

This work is a representative, but not exhaustive, treatment of recent literature on the application of impedance/dielectric spectroscopy (I/DS) for the study of nanoscale electroceramics. In this paper, the "nano" moniker is reserved for materials with grain sizes in the 10 to 100 nm range. Due to time and space limitations, the scope will be limited to non-ferroelectric materials. In ferroelectrics, much larger dielectric constants are encountered, and domain structure and dynamics play as important roles as the grain boundary-controlled behavior described in the present work. We will also not be considering nanocomposites, e.g., dispersed ionic conductors (DICs), where a finely divided (possibly nanoscale) insulating dispersed phase is added to enhance the conductivity of the matrix through space charge effects. [1,2] Although there has been outstanding recent work on nanoceramics in thin film form [3-7], our intention is to derive meaningful dielectric as well as resistivity information from I/DS. We will show that in-plane capacitance measurements on thin films will most likely be dominated by parallel apparatus and substrate capacitances. Therefore, we will consider nanoceramics in bulk form only. Finally, we will restrict our attention to materials where the local grain boundary resistivity is significantly larger than the grain core resistitivity, thus allowing for analysis by means of the widely recognized "brick layer" model (BLM), albeit with certain modifications (see below).

There is a small, but growing, literature regarding impedance spectroscopy of nanoscale electroceramics, including work on ZnO [8-10] and TiO_2 [11-13]. The largest body of work, however, has been on undoped and doped CeO_2

[13-18]. In most of these studies, two overlapping arcs are observed in Nyquist plots of impedance (Z) data ($-Z_{im}$ vs. Z_{re}). This behavior is reminiscent of the behavior in conventional electroceramics, with separate arcs for grain interiors (high frequency) and grain boundaries (low frequency). It should be stressed that it is often difficult to compare the I/DS results between separate studies on the same material, due to differences in their impurity levels and/or their microstructures. For example, some groups studied nearly dense hot-pressed specimens [12-14,17,18]; others examined partially sintered samples [11,15,16]; and still others investigated pressed compacts of nanopowders [9,10]. Nevertheless, upon careful analysis, certain trends emerge from all these studies regarding local grain boundary properties (resistivity, dielectric constant) and how these differ from those of conventional microcrystalline ceramics.

A useful framework for analysis of I/DS data is the so-called "brick layer" model (BLM), originally developed for conventional electroceramics in the 1970s and 1980s [19-22], but recently extended to the nanoscale regime [16] and to "real" microstructures [23], i.e., taking into account the actual shapes of grains. In the more conventional BLM (Fig. 1), cubic grains (core size D)

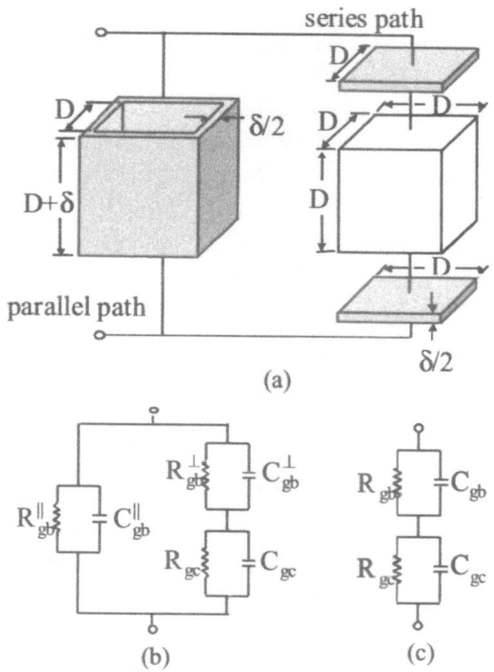

Figure 1: (a) Brick layer model with cubic grains of dimension D surrounded by grain boundaries of width δ which can be modeled by the equivalent circuit in (b), which may be simplified to the equivalent circuit model in (c).

are surrounded by layers (of boundary width δ) such that the grain size (D') is the sum of D and δ as shown. With a voltage gradient from top to bottom in Fig. 1a and grain boundaries that are more resistive than the grain cores, the equivalent circuit in Fig. 1b can be employed. Usually, the parallel grain boundary path (along the side walls) is ignored (due to their finite dimensions and high resistivity as assumed here), leaving the series path through the grain core and top and bottom grain boundaries. This leads to the simplified equivalent circuit of two parallel RC elements (one set for grain cores and one set for grain boundaries) in series with one another as in Fig. 1c. We will proceed with this simplified model throughout much of this paper, returning later to the modifications required at the finest grain sizes, i.e., to take into account the side wall contributions as D and δ become comparable in magnitude.

EXPERIMENTAL ISSUES

Before meaningful analysis of I/DS data can be made, care must be taken to ensure that the data truly reflect the sample under study and are not being influenced by stray imittances from various sources, e.g., cabling, sample holder, electrodes, etc. [24,25]. Although one wishes to measure the impedance response of the sample directly, as in Fig. 2a, this can only be done at the front panel of the impedance analyzer and only if "perfect" (low impedance) electrodes are employed. In reality, impedance measurements are made at some distance from the impedance analyzer, usually in an experimental furnace used

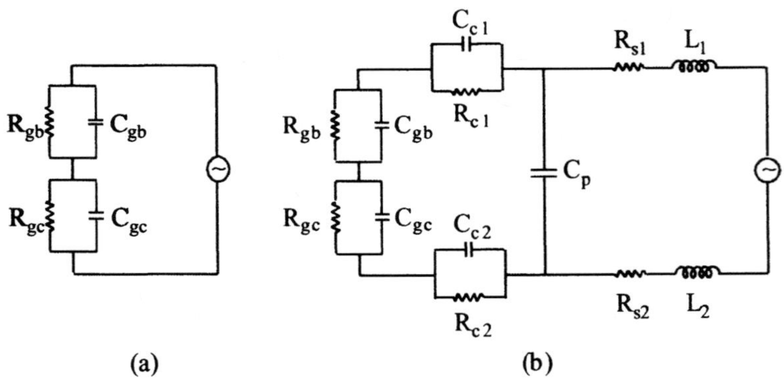

Figure 2: (a) Equivalent model of a sample with $(RC)_{gb}$ in series with $(RC)_{gc}$. The actual experimental setup is better represented by (b), in which there are contributions due to contacts (c1,c2), parallel apparatus capacitance (C_p), lead resistance (S_1,S_2) and inductance effects (L_1,L_2).

to control temperature and atmosphere. The situation is better represented by the equivalent cicuit in Fig. 2b, where series resistances (R_{s1}, R_{s2}), inductances (L_1, L_2) and parallel capacitance (C_p) contributions arise from the cabling and sample holder employed. It is possible to gain an appreciation of these contributions by performing "closed circuit" (inserting a short instead of the specimen) and "open circuit" (removing the specimen) tests, respectively. There are also protocols for null-correcting experimental data in an attempt to remove these contributions [26]. In reality, one can never fully correct the experimental data if the parallel capacitance of the apparatus is comparable to or exceeds that of the specimen. For example, unshielded alumina thermocouple protection tubes used to pass leads in and out of an experimental furnace can contribute ~ 10 pF of capacitance (per half meter of tube length). By running each I/DS lead in its own Pt-coated and grounded thermocouple protection tube, essentially extending the "coaxial cable" into the furnace, the open circuit capacitance can be lowered into the sub-pF range.

The open-circuit capacitance of the apparatus therefore sets the lower limit for capacitance measurements. This is why in-plane capacitance measurements on thin films are not tractable. If we consider a thin film 1 μm thick and 1 cm by 1 cm, the in-plane capacitance (assuming a dielectric constant of ~ 10) will be on the order of 0.0001 pF. Even with a grain boundary enhancement factor of 10^3 (see below), the effective capacitance would only be in the 0.1 pF range, i.e., below the open circuit capacitance of the best shielded/grounded system. Furthermore, a film is typically deposited on a mm-thick substrate which will contribute ~ 0.1 pF capacitance (assuming a dielectric constant of ~ 10) in parallel with the film capacitance. Therefore, in-plane I/DS measurements on thin films will register the resistance of the film (on an insulating substrate) but most likely the parallel capacitances of the apparatus and substrate. Of course, through-film (sandwich geometry) I/DS measurements do not suffer from these limitations.

Another factor to be considered in the collection and interpretation of I/DS measurements is proper electroding [24]. Fig. 3 shows the effect of imperfect electrodes on the impedance spectra of nanoscale CeO_2. What would appear to be a dual arc (grain core, grain boundary) situation actually arises from imperfect electrode contact with the sample. One or both of R_{c1} and R_{c2} (the contact resistances) in Fig. 2b become larger than the sample resistance, due to spreading resistance (current bunching) at a limited number of point contacts with the electrode. There is also a "gap" capacitance (one possible contribution to C_{c1} and C_{c2}) associated with the air-filled region between the electrode and the rough surface of the specimen. This was clearly demonstrated in that the larger (low frequency) arc could be made to disappear upon

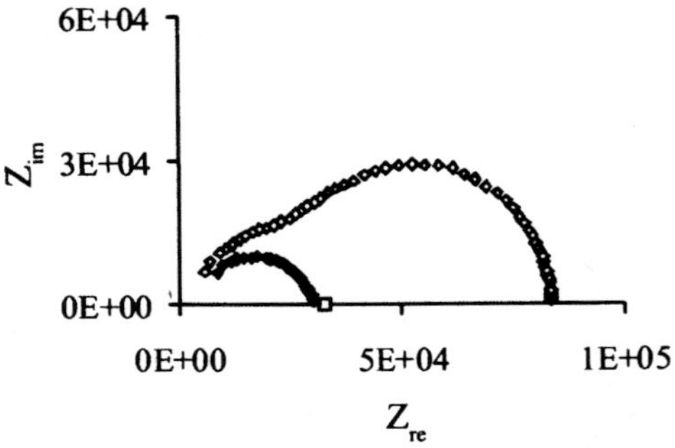

Figure 3: Nyquist plot showing the effects of imperfect electrodes on the impedance spectra of nanoscale CeO_2. Large 'dual-arc' behavior (◇) has been shown to disappear upon application of proper electroding techniques. A properly electroded sample spectrum (◆) agrees well with the 4 point DC conductivity value (□).

the application of pressure (increasing contact area) or with a better job of sputtering (filling the gap with electrode material). When properly electroded, the low frequency (rightmost) intercept agreed with the 4-point DC resistance [24]. It is strongly suggested that separate 4-point DC conductivity measurements be performed to confirm the absence of such contact effects. When measuring ionic conductors, of course, there will be additional electrode elements (not shown in Fig. 2b) and a distinct electrode arc at low frequencies arising from polarization effects at the electrode interfaces.

It is unfortunate when practitioners (including the authors!) neglect to provide frequency markers on Nyquist plots ($-Z_{im}$ vs. Z_{re}) of impedance data. Furthermore, capacitance Bode plots (real capacitance vs. frequency) are seldom supplied. To date little or no attention has been paid to the capacitive/dielectric properties of nanoscale electroceramics. Although it is possible to determine relative capacitances due to grain boundaries and grain cores from unlabeled Nyquist plots (see below), it is desirable to supply frequency markers (or at least the frequency at the top of each Nyquist arc) and C-Bode plots, if meaningful dielectric analyses are to be made.

METHODOLOGY

The premise of the current work is that good estimates of effective resistivities (for grain cores, grain boundaries) and the ratio of effective capacitances (grain boundary-to-grain core) can be obtained from Nyquist plots with well-resolved dual arcs, whether or not frequency markers have been provided. The extent of overlap of the two arcs or the "shape fit" of the overall spectrum can be used to estimate these parameters. Of course, the situation improves markedly if the frequency markers have been provided and can be accurately matched in fitting, thereby providing individual values for the grain and grain boundary capacitances. The best situation is to carry out equivalent circuit fitting of the actual raw data when this is available.

For dual-arc analysis of Nyquist data, it will be assumed that the extent of arc depression (measured as the angle from the origin through the center of the fitting semi-circle) is small or negligible, i.e., on the order of 10° or less. This is true enough in many of the abovementioned nanoceramic studies [8-10,14,16,17]. This may indicate that the microstructures of these materials (grain size and grain boundary width) are quite uniform, such that there is not a broad spread of time constants. The analysis is also valid when both arcs are depressed by similar amounts, i.e., they can each be fitted as "constant phase elements", whose complex admittances can be represented by [27]:

$$Y^{*} = Y_{o}(j\omega)^{n} \tag{1}$$

where ω is angular frequency and n is 1 for a perfect capacitor (and $Y_o = C$), but less than 1 in the case of arc depression. When both arcs have the same value of n (identical arc depression), their Y_o terms scale in the same way as their capacitances would without any arc depression. We have found that shape-fitting without arc-depression gives virtually the same ratio of capacitances. The problematic situation is where one arc (e.g., grain boundary) is significantly more depressed than the other arc (e.g., grain core) [18]. In such a case, shape-fitting without arc depression can only approximate the capacitance ratio. One would have to examine the capacitance Bode plot (real capacitance vs. frequency) to get a better picture of the individual effective capacitances.

Two situations have been simulated in Figs. 4 and 5, based upon the simple circuit in Fig. 2a. The first assumes a ratio of R_{gb}/R_{gc} of 10 and the second assumes the inverse, i.e., $R_{gb}/R_{gc} = 1/10$. The individual values selected for the components ($10^4\ \Omega$ and $10^5\ \Omega$) are arbitrary, but are not inconsistent with experimental values. A value of 10^{-11} F is assumed for C_{gc}. Again, this is arbitrary but not inconsistent with experimental values. To generate the Nyquist plots in Figs. 4a-f and Figs. 5a-f, the value of C_{gb} was systematically varied over several orders of magnitude. In each case, a perfect semi-circle is obtained when the time constants are identical, i.e., $R_{gb}C_{gb} = R_{gc}C_{gc}$. With this as the reference point, the value of $\Delta(RC)$ on each plot refers to the logarithm of the ratio of the grain boundary RC time constants in each case to that of the reference point, i.e., $\log[(R_{gb}C_{gb})/(R_{gb}C_{gb})_{ref}]$. Also given is the value of $\Delta(C)$ in each case, which is the logarithm of the ratio of the effective grain boundary

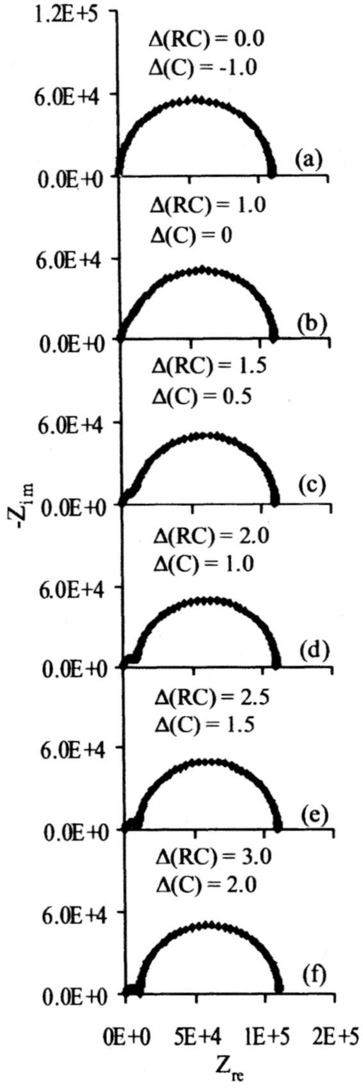

Figure 4: Nyquist plots where $R_{gb}/R_{gc} = 10$, $C_{gc} = 10^{-11}$ F, and $-1 \leq \Delta(C) \leq 2$. Arc separation is apparent when $\Delta(RC) \geq 1.0$.

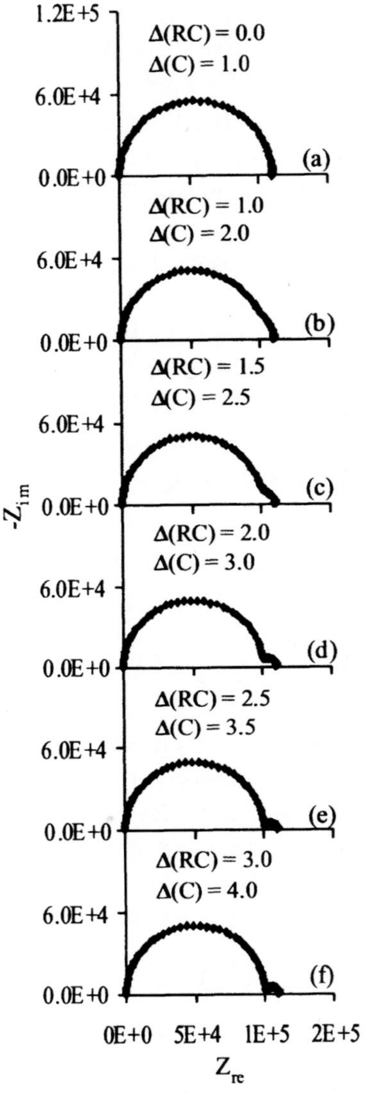

Figure 5: Nyquist plots where $R_{gb}/R_{gc} = 0.1$ $C_{gc} = 10^{-11}$ F, and $1 \leq \Delta(C) \leq 4$. Arc separation is apparent when $\Delta(RC) \geq 1.0$.

capacitance to that of the grain cores. Note that there is no plot for $\Delta(RC) = 0.5$ in each case; there was no clear indication of the second arc in this instance.

In Figs. 4 and 5 it is clear that for $\Delta(RC) \geq 1$ there is sufficient discrimination of the two arcs to estimate the ratio of R_{gb}/R_{gc} as ~ 10 or 1/10, respectively. Furthermore, it can be seen just how sensitive the overall shape is to the value of RC. At $\Delta(RC) = 1$, the grain boundary arc is just distinguishable. At $\Delta(RC) = 2$, the smaller of the two arcs is almost horizontal at its intersection with the larger arc. At $\Delta(RC) = 3$, the cusp between the two arcs almost reaches the real axis. This suggests that it is possible to obtain a reliable estimate for the effective grain boundary capacitance (or at least the ratio of the grain boundary-to-grain core capacitances) from the shape of an experimental Nyquist plot. It should be pointed out that above $\Delta(RC) = 3$, there would be little further change in the Nyquist plot. Unless frequency markers are provided, the ratio of effective capacitances could not be established by shape-fitting of Nyquist plots and must be obtained from a C-Bode plot. However, the ranges of $-1 \leq \Delta(C) \leq 2$ in Fig. 4 and $1 \leq \Delta(C) \leq 4$ in Fig. 5 cover a wide range of C_{gb}/C_{gc} values, such that much larger values would not seem to be physically reasonable. (A value of $C_{gb}/C_{gc} < 1$, i.e., $\Delta(C) < 0$, is also unrealistic, as described below.)

This kind of Nyquist shape-fit analysis is only possible if the resistance ratio, R_{gb}/R_{gc}, is between 1/10 on the low side (Fig. 5) and 10 on the high side (Fig. 4). Fortunately, most of the available nanoceramic data fall in this range [9-11,14-18]. For R_{gb}/R_{gc} values between 1/10 and 10, the analysis of Fig. 4 and Fig. 5 must be repeated. This can be carried out using the simulation mode of the "Equivalent Circuit" software [28], or by a spreadsheet routine based on the equations given in the Appendix. For R_{gb}/R_{gc} values much less than 1/10 or much larger than 10, the analysis is not recommended. It becomes increasingly difficult to accurately estimate the smaller of the two resistance values, and shape-fitting to such a small arc becomes increasingly difficult. In such cases, it is advisable to use C-Bode plots to obtain reliable capacitance information.

In certain instances, only one arc has been reported in Nyquist plots of nanoceramics [8,12,13]. One possible explanation is that the two microstructural regions possess identical time constants, i.e., $R_{gb}C_{gb} = R_{gc}C_{gc}$. This is certainly possible if $R_{gb} < R_{gc}$ (requiring $C_{gb} > C_{gc}$), but is unlikely for $R_{gb} > R_{gc}$ (requiring $C_{gb} < C_{gc}$). In the brick layer model, the ratio of effective capacitances is given by [27]:

$$C_{gb}/C_{gc} = \left(\varepsilon_{gb}/\varepsilon_{gc}\right)\left(D/\delta\right) \tag{2}$$

where ε is dielectric constant, D is grain size, and δ is grain boundary width. Both factors on the right side of this equation are expected to increase C_{gb} relative to C_{gc}, given that D is typically larger than δ, and that grain boundary regions tend to be more polarizable than grain core regions (see below). A more likely scenario is that one resistance proves to be orders of magnitude larger than the other resistance. In such a case the lower resistance arc will not be visible in the Nyquist plot. If the remaining arc is negligibly depressed below the real axis, its capacitance can be determined according to [27]:

$$C = (\omega_{top} R)^{-1} \tag{3}$$

where R is the diameter of the arc (or the resistance of the element), $\omega_{top} = 2\pi\nu_{top}$, and ν_{top} is the frequency at the top of the arc. This capacitance can, in turn, be compared with the bulk (grain core) capacitance calculated from the published single crystal dielectric constant and the measured geometry of the specimen. If it is much greater than the bulk capacitance, it probably arises from the grain boundaries, and $R_{gb} >> R_{gc}$. If the two capacitances are comparable, the arc is probably attributable to the grain cores ($R_{gb} << R_{gc}$).

ANALYSIS AND DISCUSSION

Grain Boundary Resistance

Using shape-fit analysis of dual-arc Nyquist data or, in certain cases, RC analysis of single-arc Nyquist plots (using Eq. 3), grain boundary resistance information was calculated from various literature sources. Effective resistivities were calculated based upon as-pressed specimen geometries:

$$R_{gb} = \rho_{gb}\ (A/l) \tag{4}$$

where ρ_{gb} is the effective grain boundary resistivity, R_{gb} is the experimental grain boundary resistance (i.e., the diameter of the grain boundary arc), and A/l represent the overall sample dimensions, i.e., cross-sectional area (A) and interelectrode spacing (l). The as-pressed specimen dimensions were employed in each case, since few studies report final (as-fired) dimensions. This should not significantly alter the magnitude of the results.

It is of interest to compare the grain boundary resistances between different studies on the same nanoceramic, and also between nanoscale and microscale ceramics. A common approach is to calculate "specific" (local) grain boundary resistivity (ρ_{sgb}) from the effective grain boundary resistivity, using [27]:

$$\rho_{sgb} = \rho_{gb}\ (D/\delta) \tag{5}$$

However, this requires an independent knowledge of the grain boundary width (δ). Instead, we have opted to calculate the areal grain boundary resistance (the resistance of a cm^2 unit area of grain boundary, R_{agb}) [29]:

$$R_{agb} = \rho_{gb} / n_{gb} \tag{6}$$

where n_{gb} is the grain boundary density (number of grain boundaries per unit length). For an equiaxed microstructure with a narrow distribution of grain sizes, n_{gb} is essentially the inverse of the grain size ($n_{gb} \sim 1/D'$). The results of this analysis are given in Table 1.

There are two important observations to be made from the resistance information in Table 1. First, there are fairly wide variations in grain boundary resistance for the same material having similar grain size but made by different means, i.e., having different impurity contents. Second, it is obvious that, with the exception of one TiO_2 study [12], the areal grain boundary resistance tends to increase by 3-5 orders of magnitude when the grains are grown from the 10-20 nanometer range into the micrometer range. This means that individual grain boundaries are far less resistive in nanoceramics than in the corresponding microcrystalline materials. Many of the literature studies cited in the Introduction have reached the same conclusion.

Chiang has discussed several possible explanations for the grain size-dependent resistance of grain boundaries in nanoceramics [30]. At sufficiently small grain sizes, nanocrystalline materials can be dominated by the defect properties of their interfaces. Several groups have demonstrated that the reduction enthalpy associated with oxygen vacancy formation in nanocrystalline CeO_2 is less than half the bulk value [14,15,17,18]. When the vacancy concentration is fixed by acceptor impurities, intentional or otherwise, the electron population can be orders of magnitude higher (and the local resistivity orders of magnitude lower) than in a correspondingly doped microcrystalline material. Another possibility involves impingement and overlap of space-charge regions, i.e., when the grain size approaches the scale of the Debye length. This can alter both the grain boundary "width" and the height of potential barriers at grain boundaries [2]. Finally, grain size-dependent solute segregation must be taken into account [13,30]. The specific surface area of grain boundaries increases dramatically as grain size is reduced into the nanometer range. The amount of solute competing for energetically favorable grain boundary sites becomes limited. This will influence both the height of grain boundary barriers and the extent of space charge regions adjacent to the boundaries.

Table 1: Analysis of impedance/dielectric spectra

Citation	Oxide (Temp, °C)*	Grain size (nm)	log R_{agb} (nano)†	log R_{agb} (micro)†	C_{gb}/C_{gc}‡
8	ZnO (500)	60	0.4	-	5-10**
9	ZnO (100)	10	-2.9	-	= 10^3
12	TiO_2 (580)	35	0.4	1.1 (1μm)§	~ 1**
11	TiO_2 (156)	10-20	-0.1-0.2	4.9-5.2 (10-20μm)§	~10^3
14	CeO_2 (400)‖	10	0.2	3.5 (5μm)§	~ 1
14	CeO_2 (400)¶	10	-2.3	2.7 (5μm)§	~ 100
16	CeO_2 (500)	18	-1.2	2.6 (3 μm)§	~ 10^3
17	Gd-doped CeO_2 (450)	10	-0.4	-	~ 30
18	CeO_2 (460)	26	0.5	5.83 (10 μm)§	~ 30

* Measurement temperature
† Areal grain boundary resistance (see text)
‡ Ratio of grain boundary-to-grain core capacitance
** C_{gb} determined by single-arc method (see text)
§ Grain size of microcrystalline specimen
‖ Powder made by inert gas condensation
¶ Powder made by chem-prep technique

Grain Boundary Capacitance

Few prior works have dealt with the issue of grain boundary capacitance in nanoceramic materials. Again, we have opted *not* to consider "specific" grain boundary capacitance, C_{sgb} [30],

$$C_{sgb} = C_{gb}\,(\delta/D) \quad (7)$$

which requires independent knowledge of the grain boundary "width" (δ). Instead, by shape-fitting of Nyquist dual-arc data or RC analysis of single-arc data (Eq. 3), we can estimate the ratio of the effective capacitances due to grain boundaries and grain cores (C_{gb}/C_{gc}). When frequency markers were provided on Nyquist plots, we were able to confirm that the grain core capacitances were in reasonable agreement with published values of dielectric constant [31,32], taking into account the reported specimen geometries. The results for various literature studies are given in Table 1.

In one case, shape-fitting was not possible, due to the grain boundary arc appearing as an ill-defined shoulder (with $R_{gb} \sim 1/10\,R_{gc}$) at low frequencies in Nyquist plots [16]. Fortunately, capacitance Bode plots were provided in that study. Figs. 6 and 7 show real capacitance Bode plots corresponding to the Nyquist plots in Figs. 4 and 5. As before, the component resistance values were fixed as $10^4\,\Omega$ and $10^5\,\Omega$, and the grain core capacitance was fixed as 10^{-11} F. The effective grain boundary capacitance was then varied over a wide range, $-1 \leq \Delta(C) \leq 2$ in Fig. 6 and $1 \leq \Delta(C) \leq 4$ in Fig. 7, where $\Delta(C)$ is the logarithm of the ratio of the effective grain boundary capacitance to that of the grain cores, i.e., $\log\,(C_{gb}/C_{gc})$. Also shown in each figure is a plot of $\log[C_{re}(\mathrm{lo}\nu)/C_{re}(\mathrm{hi}\nu)]$ vs. $\log\,(C_{gb}/C_{gc})$. Each real capacitance Bode plot shows two plateaus, one at low frequency ($C_{re}(\mathrm{lo}\nu)$) and one at high frequency ($C_{re}(\mathrm{hi}\nu)$). Given either situation, $R_{gb} \sim 10R_{gc}$ (Fig. 6) or $R_{gb} \sim 1/10\,R_{gc}$ (Fig. 7), the effective capacitance ratio can be estimated from Fig. 6b or Fig. 7b, respectively, based upon observable plateaus in the experimental C-Bode data. In this way, a value of $\log\,(C_{gb}/C_{gc})$ was obtained for the data in Ref. [16]. It should be stressed that situations with resistance ratios between $R_{gb}/R_{gc} = 1/10$ and $R_{gb}/R_{gc} \sim 10$ must be individually modeled using "Equivalent Circuit" or the equations in the Appendix. Extracting capacitance ratios for resistance ratios outside this range is not advisable unless high quality Z-Bode and C-Bode data are available, with clear dual-plateau behavior [16].

What is immediately obvious from the capacitance ratio data in Table 1 is that a wide range of values are obtained, from essentially unity to in excess of 10^3, even with the same host material and virtually identical grain sizes. This is

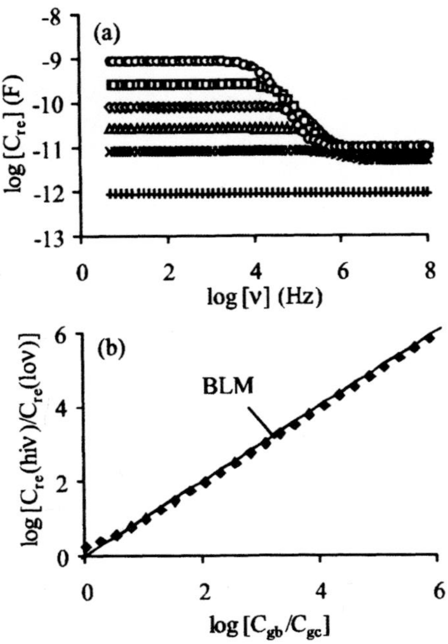

Figure 6: (a) Real capacitance Bode plot where $R_{gb}/R_{gc} = 10$ and $C_{gc} = 10^{-11}$ F. C_{gb} varies so that [Δ(RC),Δ(C)] = O:[3,2]; □:[2.5,1.5]; ◇:[2,1]; △:[1.5,0.5]; x:[1,0] +:[0,-1]. (b) The effective capacitance ratio versus actual capacitance ratio.

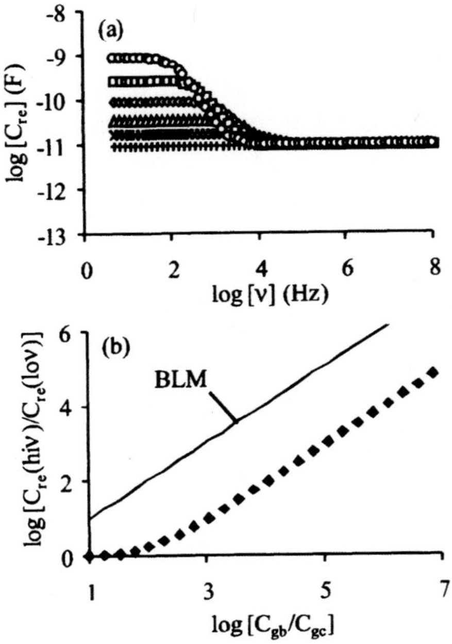

Figure 7: (a) Real capacitance Bode plot where $R_{gb}/R_{gc} = 10$ and $C_{gc} = 10^{-11}$ F. C_{gb} varies so that [Δ(RC),Δ(C)] = O:[3,4]; □:[2.5,3.5]; ◇:[2,3]; △:[1.5,2.5]; x:[1,2]; +:[0,1]. (b) The effective capacitance ratio versus actual capacitance ratio.

most likely indicative of variations in overall, as well as local, impurity concentrations from sample to sample.

Small capacitance ratios in Table 1 are relatively easy to explain in terms of Eq. 2. Grain size and grain boundary "width" may be comparable in these samples, due to broad space charge regions. At the same time, the dielectric constant in these regions may not be significantly different from bulk (and residual grain core) regions. In instances when space charge regions should be insignificant, e.g. in highly doped materials, a more complicated analysis may be required [12].

On the other hand, large capacitance ratios are more difficult to understand. In at least four instances, the grain boundary arc appeared as a small feature at low frequencies (on the right side) of the dominant grain core arc in Nyquist plots [9,11,14,16]. For such a feature to appear, the effective grain boundary capacitance must be much larger than the effective grain core capacitance, given that the grain boundaries are much less resistive than the grain cores. For example, to achieve $\Delta(RC) = 1$ in the $R_{gb} = 1/10\ R_{gc}$ situation requires a $\Delta(C)$ of 2, i.e., C_{gb} two orders of magnitude larger than C_{gc}. This follows from the mathematical analysis in the Appendix. For $R_{gb} << R_{gc}$ it is shown that:

$$C_{re}(\text{lo}\nu) / C_{re}(\text{hi}\nu) \rightarrow 1 + \left(R_{gb} / R_{gc}\right)^2 \left(C_{gb} / C_{gc}\right) \tag{8}$$

The $(R_{gb}/R_{gc})^2$ term in this equation accounts for the simulated data lying approximately two orders of magnitude or $(1/10)^2$ below the line representing $[C_{re}(\text{lo}\nu)/C_{re}(\text{hi}\nu)] = (C_{gb}/C_{gc})$ in Fig. 7b. A large grain boundary capacitance was confirmed in brick layer model (BLM) fitting of C-Bode data in Ref. [16].

To account for capacitance ratios in the $10^2 \leq C_{gb}/C_{gc} \leq 10^3$ range or larger within the BLM framework, it is necessary to combine large ratios of (D/δ) and $\varepsilon_{gb}/\varepsilon_{gc}$. There is a practical upper limit to $(D/\delta) \sim 100$ for D' ~ 10 nm, since δ cannot be smaller than atomic dimensions (~ 0.1 nm). It is more likely that δ is on the nanometer scale. In either case, capacitance ratios larger than 10^2 require substantially higher local dielectric constants at grain boundaries. This may be a result of highly disordered and more polarizable regions near grain boundaries, but may also be a function of solute segregation to grain boundaries, which can also influence the local polarizability. Differences in solute segregation to grain boundaries arising from different processing and/or overall impurity content may also account for sample-to-sample variations for measurements on otherwise identical materials (same oxide and comparable grain size).

Modified Brick Layer Modeling

We began this review by assuming that the local (or specific) grain boundary resistivity was greater than grain core resistivity, in order to apply the brick layer model. It is interesting to consider how this can remain true, while achieving effective grain boundary resistances (and resistivities) smaller than those of the grain cores. Elsewhere [16] we have developed the equations to model the frequency-dependent impedance/dielectric behavior of the equivalent circuit in Fig. 1b. This modified BLM assumes the cubic grain geometry of Fig. 1a and takes into account the side-wall contributions. Each region, grain core and grain boundary, has a fixed specific resistivity. For simplicity, it is assumed that local grain boundary and grain core dielectric constants are identical.

Figure 8a shows how the ratio of $C_{re}(lo\nu)/C_{re}(hi\nu)$, taken from the two plateaus in C-real Bode plots, varies with grain size for three different values of resistivity ratio. Here, the resistivity ratio is the ratio of the local or specific resistivity of grain boundaries to that of the grain cores, i.e., ρ_{sgb}/ρ_{sgc}. Each curve goes through a maximum where the resistivity ratio is equal to the ratio, D/δ. At this point the two contributions have equal effective resistances, $R_{gb} = R_{gc}$. For example, the grain boundaries are 10^3 times more resistive than the grain cores when $\rho_{sgb}/\rho_{sgc} = 1000$, however at $D/\delta = 10^3$ grain boundaries comprise only 1/1000th of the series path in Fig. 2c, so the resistances are equal.

Our prior work showed that at grain sizes less than the maximum, the corresponding Nyquist plots look like those in Fig. 4, with the grain boundary (low frequency, rightmost) arc being larger than that of the grain cores. This is the classic BLM regime, and it can be seen that $C_{re}(lo\nu)/C_{re}(hi\nu)$ scales directly with D/δ in accordance with Eq. 2. For example, $C_{re}(lo\nu)/C_{re}(hi\nu)$ increases from 10^1 to 10^2 as D/δ increases from 10^1 to 10^2 (for a resistivity ratio of 1000).

At grain sizes larger than the peak in Fig. 8a, there is a decrease in $C_{re}(lo\nu)/C_{re}(hi\nu)$. This follows from Eq. 7, as described previously, and is due to the effective grain boundary resistance becoming less than that of the grain cores. The Nyquist plots were identical to those in Fig. 5, with the grain boundary (low frequency, rightmost) arc being much smaller than the grain core arc. Although the resistivity ratio is still such that grain boundaries represent a local barrier to charge transport, at large values of D/δ there are too few grain boundaries in the series path of Fig. 1b to generate a large grain boundary resistance.

The analysis in Ref. [16] indicated that impedance data for nanoceria fell predominantly to the right of the peak in Fig. 8a, and that the resistivity ratios required for fitting were quite small, in the 1-40 range. However, when grain growth into the micrometer range was carried out, the situation reverted to the classic BLM situation, with the data falling to the left of the peak, and with a much larger resistivity ratio ($\sim 10^4$). This analysis is consistent with the present results indicating that the areal grain boundary resistance is much lower in nanophase ceramics than in their conventional microcrystalline counterparts.

An additional observation is that data must fall at or near a peak in Fig. 8a for two distinct arcs to be observed in Nyquist plots. At the very peak, two arcs of equal size are obtained [16]. As D/δ is increased or decreased by an order of magnitude, we reach the $R_{gb} = 1/10\ R_{gc}$ or $R_{gb} = 10R_{gc}$ situations, respectively. The most interesting aspect of the nanoceramic I/DS data considered in this study is that two arcs are usually observed. This supports the notion

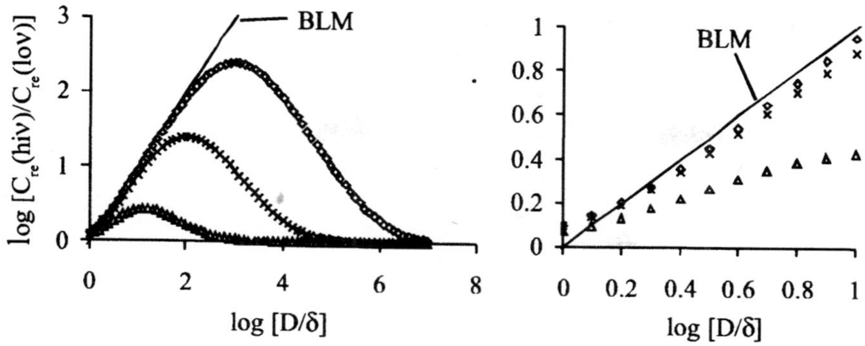

Figure 8: (a) Effects of grain size on C_{re}(lov)/C_{re}(hiv), where R_{gb}/R_{gc} = 1000(◇); 100(×); and 10(△). (b) Enlargement showing the small grain size region of (a).

that the local (specific) grain boundary resistivities must be relatively low in nano- vs. micro-scale ceramics.

It is interesting to consider the role of side-wall contributions to the impedance/dielectric behavior. Figure 8b shows an enlargement of the small grain size region of Fig. 8a. The deviation from BLM behavior at larger D/δ values is due to the nearby maximum in Fig. 8a, and is not associated with side-wall effects. As the resistivity ratio increases and the maximum moves to larger values of D/δ, the curves approach the classical BLM line. However, all three curves fall above the BLM line as $D/\delta \rightarrow 1$ ($\log[D/\delta] \rightarrow 0$), as shown in Fig. 8b. This is due to the side-wall contributions, which must therefore be taken into account whenever a combination of small grain size and large grain boundary "width" is encountered.

CONCLUSIONS

With proper attention to cabling, electroding, and correction for stray apparatus imittances, impedance/dielectric spectroscopy is a powerful tool to study the frequency-dependent electrical/dielectric properties of nanoceramics. The present work outlined a technique to evaluate the effective resistances and capacitances due to grain cores and grain boundaries in conventional (non-ferroelectric) ceramics, based upon shape-fitting of dual-arc Nyquist plots, whether or not frequency markers have been supplied. When possible, however, this procedure should be supplemented by standard C-Bode plot analysis and overall equivalent circuit fitting to the raw data.

The available literature for bulk nanoceramics (ZnO, CeO_2, TiO_2) indicates that the areal grain boundary resistance (the resistance of a single grain boundary of unit area) tends to be 3-5 orders of magnitude smaller in nanoceramics than in their microcrystalline counterparts. This is most likely associated with grain size-dependent phenomena, e.g., impurity segregation to grain boundaries and/or decreased defect formation energies at grain boundaries. There is considerable variation in areal grain boundary resistance between specimens of the same material and having the same grain size, but with different impurity contents.

The capacitance ratio, grain boundary-to-grain core, in a given nanoceramic material was found to vary considerably from sample to sample, even at the same grain size. This was found by both shape-fitting of dual-arc Nyquist plots and by real capacitance Bode plot analysis. In several instances, large values of the ratio, C_{gb}/C_{gc}, were obtained, some in excess of 10^3. At such small grain sizes, this would require both small values of grain boundary width and large local dielectric constants, according to conventional brick layer model (BLM) analysis.

A modified BLM analysis was employed to account for the unique dual-arc behavior seen in the Nyquist plots of nanoceramic specimens. There is a peak in the capacitance ratio, $C_{rc}(\text{lo}\nu)/C_{rc}(\text{hi}\nu)$, vs. D/δ plot where $R_{gb} = R_{gc}$. At smaller grain sizes, the effective grain boundary resistance is larger than that of the grain cores, and conventional BLM behavior is observed. However, at grain sizes larger than the peak, the effective grain boundary resistance becomes smaller than that of the grain cores. The resulting $C_{rc}(\text{lo}\nu)/C_{rc}(\text{hi}\nu)$ behavior will be substantially less than the ratio of C_{gb}/C_{gc} in such an instance.

The BLM does not take into account side-wall contributions to the electrical/dielectric properties. A modified BLM has been developed, which accounts for side-wall contributions and predicts the behavior as D and δ become comparable in size.

APPENDIX

The equivalent circuit consisting of two RC elements in series as shown in Fig. 1c can be described by a single impedance value of the form $R + jX$. The impedance of a resistor and capacitor element are R and $(j\omega C)^{-1}$ respectively where j denotes $\sqrt{-1}$, ω is the angular frequency ($2\pi\nu$), and R and C are the resistance and capacitance, respectively. Combining the resistance and capacitance impedances in parallel, each RC element can be replaced by an equivalent impedance value given below:

$$Z_i = \frac{R_i (j\omega C_i)^{-1}}{R_i + (j\omega C_i)^{-1}} = \frac{R_i - jR_i^2 \omega C_i}{(R_i C_i \omega)^2 + 1} \tag{A1}$$

where i denotes gb or gc. The total impedance of the system (Z_{total}), therefore, is the combination of the impedances given by Eq. A1 in series ($Z_{gb} + Z_{gc}$). The real and imaginary portions of the total impedance are given below:

$$Z_{total,re} = \frac{R_{gb}}{(R_{gb}C_{gb}\omega)^2 + 1} + \frac{R_{gc}}{(R_{gc}C_{gc}\omega)^2 + 1} \tag{A2a}$$

$$Z_{total,im} = \frac{R_{gb}^2 C_{gb}\omega}{(R_{gb}C_{gb}\omega)^2 + 1} + \frac{R_{gc}^2 C_{gc}\omega}{(R_{gc}C_{gc}\omega)^2 + 1} \tag{A2b}$$

The capacitance of the system can be determined from the total equivalent impedance by the expression $C_{re} = -Z_{total,im}/(\omega|Z_{total}|^2)$ [16], where $|Z_{total}|^2$ is the modulus squared of the impedance and is given by $Z_{total,re}^2 + Z_{total,im}^2$. Using Eq. A2a and b, the real component of the capacitance is given by:

$$C_{re} = \frac{R_{gb}^2 C_{gb}\left[(R_{gc}C_{gc}\omega)^2 + 1\right] + R_{gc}^2 C_{gc}\left[(R_{gb}C_{gb}\omega)^2 + 1\right]}{(R_{gb}R_{gc}\omega)^2 (C_{gc} + C_{gb})^2 + (R_{gc} + R_{gb})^2} \tag{A3}$$

Two limiting cases are considered, $\omega \rightarrow \infty$ (C_{re}(hiv)) and $\omega \rightarrow 0$ (C_{re}(lov)). At high frequencies Eq. A3 simplifies to:

$$C_{re}(\text{hiv}) = C_{gb}C_{gc}/(C_{gb} + C_{gc}) \tag{A4}$$

In the limit of $\omega \rightarrow 0$, however, the resistance of both the grain boundaries and cores partially govern the capacitance of the system.

$$C_{re}(\text{lov}) = \frac{R_{gb}^2 C_{gb} + R_{gc}^2 C_{gc}}{(R_{gb} + R_{gc})^2} \tag{A5}$$

In the case where $R_{gb} >> R_{gc}$ this equation reduces to:

$$C_{re}(\text{lov}) \sim C_{gb} + (R_{gc}/R_{gb})^2 C_{gc} \tag{A6}$$

and in the opposite case where $R_{gc} >> R_{gb}$ the capacitance reduces to:

$$C_{re}(lo\nu) \sim C_{gc} + (R_{gb}/R_{gc})^2 C_{gb} \qquad (A7)$$

Combining Eq. A4 and A7 in the case where $R_{gc} >> R_{gb}$ it can be shown that the ratio of $C_{re}(lo\nu)$ to $C_{re}(hi\nu)$ when $C_{gb} >> C_{gc}$ (which is typically the case) is given by:

$$\frac{C_{re}(lo\nu)}{C_{re}(hi\nu)} = 1 + \left(\frac{R_{gb}}{R_{gc}}\right)^2 \left(\frac{C_{gb}}{C_{gc}}\right) \qquad (A8)$$

Note that in the opposite case when $R_{gb} >> R_{gc}$ and $C_{gb} >> C_{gc}$ the combination of Eq. A4 and A6 gives:

$$\frac{C_{re}(lo\nu)}{C_{re}(hi\nu)} = \frac{C_{gb}}{C_{gc}} + \left(\frac{R_{gc}}{R_{gb}}\right)^2 \qquad (A9)$$

which is approximately C_{gb}/C_{gc} as seen in Fig. 6.

ACKNOWLEDGMENTS

This work was supported in part by the NSF-MRSEC program under Grant No. DMR-0076097 and by the U.S. Dept. of Energy under Grant No. FG02-84-ER45097. The authors are grateful to Dr. Kosacki of the University of Missouri-Rolla for providing thin film data.

NOTES

* Current address: Corporate R&D Center; SAMSUNG SDI CO. LTD.; 575 Shin-Doing Paldal-Gu Suwon-City Kyungki-Do Korea 442-391

REFERENCES

1. N. J. Dudney, Ann. Rev. Mater. Sci., 19, 103 (1989).
2. J. Maier, Prog. Solid State Chem., 23, 171 (1995).
3. I. Kosacki and H. U. Anderson, Appl. Phys. Lett., 69, 4171 (1996).

4. I. Kosacki and H. U. Anderson, Solid State Ionics, 97, 429 (1997).
5. I. Kosacki and H. U. Anderson, Mat. Res. Soc. Symp. Proc., 453, 537 (1997).
6. I. Kosacki, B. Gorman and H. U. Anderson, in Ionic and Mixed Conducting Ceramics, eds. T. A. Ramanarayanan, W. L Worrell, H. L. Tuller, A. C Khandkar, M. Mogensen and W. Gopel, Proc. Electrochem. Soc., 97-24, 631 (1997).
7. I. Kosacki and H. U. Anderson, Ionics, 6, 294 (2000).
8. J. Lee, J. –H. Hwang, J. J. Mashek, T. O. Mason, A. E. Miller and R. W. Siegel, J. Mater. Res., 10, 2295 (1995).
9. J. Jose and M. A. Khadar, Nanostruct. Mater., 11, 1091 (1999).
10. J. Jose and M. A. Khadar, Mat. Sci. Eng. A, 304-306, 810 (2001).
11. S. Bhowmik, K. P. Constant, J. C. Parker and M. Ali, Mat. Sci. Eng. A, 204, 258 (1995).
12. P. Knauth and H. L. Tuller, J. Appl. Phys., 85, 897 (1999).
13. P. Knauth and H. L. Tuller, Solid State Ionics, 136-137, 1215 (2000).
14. Y. –M. Chiang, E. B. Lavik, I. Kosacki, H. L. Tuller and J. Y. Ying, J. Electroceram., 1, 7 (1997).
15. J. –H. Hwang and T. O. Mason, Z. Phys. Chem., 207, 21 (1998).
16. J. –H. Hwang, D. S. McLachlan and T. O. Mason, J. Electroceram., 3, 7 (1999).
17. Y. –M. Chiang, E. B. Lavik and D. A. Blom, Nanostruct. Mater., 9, 633 (1997).
18. A. Tschöpe, E. Sommer and R. Birringer, Solid State Ionics, 139, 255 (2001).
19. J. E. Bauerle, J. Phys. Chem. Solids, 30, 2657 (1969).
20. N. M. Beekmans and L. Heyne, Electrochem. Acta, 21, 303 (1976).
21. T. van Dijk and A. J. Burgraaf, Phys. Stat. Sol. A, 63, 229 (1981).
22. M. J. Verkerk, B. J. Middelhuis and A. J. Burgraaf, Solid State Ionics, 6, 159 (1982).
23. J. Fleig and J. Maier, J. Electrochem. Soc., 145, 2081 (1998).
24. J. –H. Hwang, K. S. Kirkpatrick, T. O. Mason and E. J. Garboczi, Solid State Ionics, 98, 93 (1997).
25. D. D. Edwards, J. –H. Hwang, S. J. Ford and T. O. Mason, Solid State Ionics, 99, 85 (1997).
26. B. J. Christensen, R. T. Coverdale, R. A. Olson, S. J. Ford, E. J. Garboczi, H. M. Jennings, and T. O. Mason, J. Am. Ceram. Soc., 77, 2789 (1994).
27. J. R. Macdonald, Impedance Spectroscopy: Emphasizing Solid Materials and Systems, Wiley & Sons, New York (1987).
28. B. A. Boukamp, "Equivalent Circuit (EQUIVCRT.PAS)", University of Twente, The Netherlands (1990).
29. M. Miyayama, H. Inoue and H. Yanagida, J. Am. Ceram. Soc., Vol., C164 (1983).
30. M. Aoki, Y. –M. Chiang, I. Kosacki, L. J. –R. Lee, H. Tuller and Y. Liu, J. Am. Ceram. Soc., 79, 1169 (1996).
31. R. J. Brook, Concise Encyclopedia of Advanced Ceramic Materials (Pergamon Press, 1991).
32. D. Y. Wang, D. S. Park, J. Griffith and A. S. Nowick, Solid State Ionics, 2, 95 (1981).

CHARACTERIZATION OF NANOCRYSTALLINE OXIDES BY EXAFS SPECTROSCOPY

Alan V. Chadwick and Georgina E. Rush
Centre for Materials Research, School of Physical Sciences, University of Kent, Canterbury, Kent CT2 7NR, UK.

I. INTRODUCTION

Nanostructured materials and nanophase systems are terms used to describe materials with structure on the scale of 1 to 100 nm. In comparison to their bulk counterparts they often have unique physical and chemical properties [1-7] and are currently the focus of intense academic and technological interest. In general terms the origins of the unusual behaviour are (i) that the particles have dimensions comparable to the length scale of basic quanta in solids (e.g. the de Broglie wavelength of electrons, phonon wavelengths, mean-free path of excitons, etc.) and/or (ii) surface effects dominate the thermodynamics and energetics of the particles (e.g. crystal structure, surface morphology, reactivity, etc.). This chapter will concentrate on one type of nanocrystalline material, namely the simple binary oxides, which have a wide range of applications, including specialised ceramics, solid electrolytes, heterogeneous catalysts and catalyst supports. Ceramics made from these materials can be both superhard, due to the dimensions being too small to support Frank-Read dislocation loops, and superplastic, due to the large number of grain boundaries. The surface reactivity of a nanocrystal would be expected to be high due to the high surface-to-volume ratio, however there seems to be an additional enhancement factor arising from unusual surface morphologies [8,9]. In ionic materials there is the prospect of novel point defect possibilities when the diameter of the nanocrystal is smaller than the width of the space-charge layer [10,11] which would result in unusually high ionic conductivities with implications for improved performance of electrolytes in fuel cells, batteries and sensors [12].

As with all nanocrystalline materials the key to understanding the unusual properties of the oxides is a detailed knowledge of their atomic level microstructure. One approach to information on the microstructure is

Extended X-ray Absorption Fine Structure (EXAFS) measurements and it is has been used by several groups in the study of nanocrystalline materials. The objective of this chapter is to present an overview the EXAFS work on the nanocrystalline oxides. In the space available it is not possible to review all the work in this area and we shall focus on the systems we have been investigating over the last decade, namely tin oxide (SnO_2), zinc oxide (ZnO), zirconium oxide (zirconia, ZrO_2) and cerium oxide (ceria, CeO_2) and the current view of the microstructure of materials. We are aware that many of the readers of this volume will not be familiar with EXAFS technique. Therefore in the next Section we will outline the technique, describing the principles of the technique, the experimental arrangement and the data analysis. References will be given to more detailed texts. This Section will be followed by a consideration by each of the materials, dealing firstly with the pure materials and following this with a discussion of the doped oxides.

Before we begin any discussion of the details of the technique and the materials it is worth reviewing the models that have been proposed for the microstructure of nanocrystals. A schematic picture that is often used to represent a nanocrystal is shown in Figure 1 and this shows the crystalline grains and the intergrain region. A fundamental question is the extent of the disordered region between the grains. In simple terms, are these normal grain boundaries like those found in bulk solids and only a few atoms thick, or are they more extensive? The thickness of the interfacial region will affect the fraction of atoms in the sample that are in surface sites and some estimates for magnesium oxide are shown in Figure 2. It is worth noting that in the early work on nanocrystals several authors intuitively assumed that there was extensive disorder in the interfacial regions and that this was a cause of many of the unusual properties of nanocrystalline materials. We will see that the

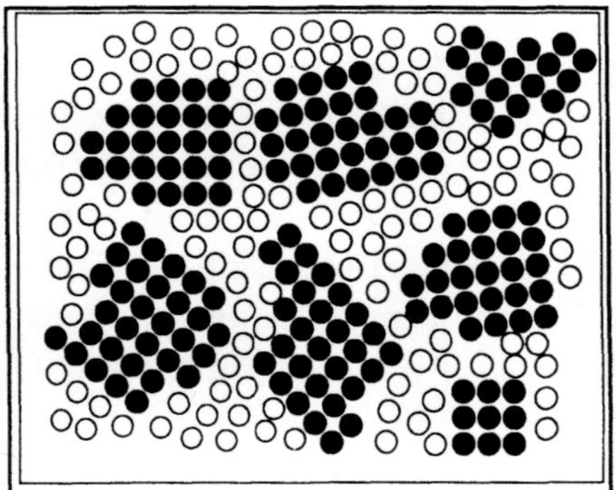

Figure 1. A two-dimensional schematic representation of a nanocrystalline material; filled circles are atoms in crystal grains, empty circles are in interfaces.

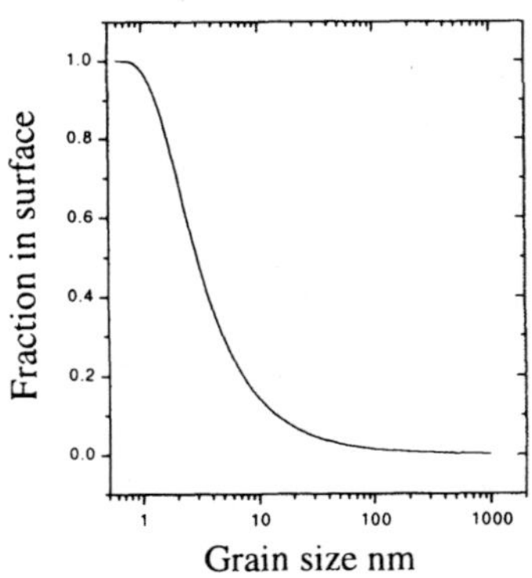

Figure 2. The fraction of atoms in the surface of a spherical grain.

resolution of this question can be achieved by EXAFS experiments. A common feature of nanocrystalline materials is their ability to dissolve large amounts of impurity atoms. The location of these atoms, in the crystal grains or in the interfaces, is another problem that can be resolved by EXAFS experiments. We will see later in this chapter how the nature of the grain boundaries and the position of impurity atoms has been determined for nanocrystalline oxides.

II. EXAFS METHODOLOGY

II.1. Basic Theory

In X-ray absorption spectroscopy the absorption coefficient, μ (= *log [incident intensity/transmitted intensity]*) of a sample is measured as a function of the incident photon energy across the absorption edge for the ejection of a core level (*K* or *L*) electron. The typical spectrum for condensed matter is shown in Figure 3. Beyond the absorption edge there are oscillations referred to as fine structure, which can be about 1/10 of the size of the edge step and which decrease in magnitude with increasing X-ray energy. The whole region is referred to as X-ray Absorption Fine Structure (*XAFS*) and is divided into two regions, the X-ray Absorption Near Edge Structure (XANES) extending about 50 eV beyond the edge, and the Extended X-ray Absorption Fine Structure (EXAFS) extending typically 1000 eV beyond the edge. In addition, there can be pre-edge features in the spectrum. All these three types of feature provide structural information about the target atom, the atom which is emitting the photoelectron. The pre-edge features arise from excitations of the core electron to higher states within the atom and can fingerprint the oxidation state of the atom. In the XANES region the photoelectron is moving slowly and becoming involved with orbitals of the neighbouring atoms. This region

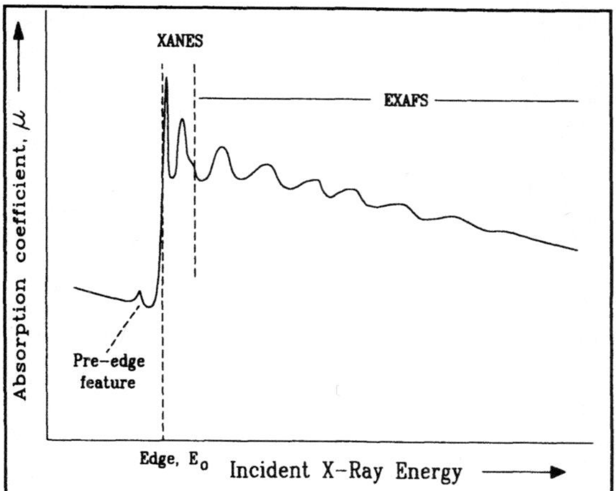

Figure 3. A typical XAS spectrum.

provides a fingerprint for the oxidation state of the atom and its local co-ordination geometry. It is the oscillations in the EXAFS region that can be analysed in detail to provide quantitative information on the local environment. In this region the photoelectron is moving rapidly and is subject to single scattering events with the surrounding atoms. The theory of EXAFS can be found in several texts [13-16]. The oscillations arise from the photoelectron wave being backscattered and interfering with the outgoing wave. If the two waves are in phase there will be constructive interference, a lower final state energy and a higher probability for absorption. If the two waves are out of phase then there will be destructive interference, higher final state energy and a lower probability for absorption. Thus as the incident photon energy increases so does the energy of the emitted photoelectron with consequential changes on its wavelength. Since the distance between the target atom and its neighbours is fixed there will be shifts in and out of phase and hence the observation of the EXAFS oscillations.

From the qualitative explanation in the preceding paragraph it should be clear that the frequency of the EXAFS oscillations contains information on the distance from the target atom from its neighbours. The intensity of the oscillations will depend on the type of atom which is acting as the backscatterer, i.e. the higher the atomic weight the more intense the oscillations, and the number of backscattering atoms. To be more precise, after subtraction of the background absorption, the normalised absorption coefficient, $\chi(k)$, as a function of the photoelectron momentum, k, can be written in an equation of the form [15,17]:-

$$\chi(k) =$$

$$\sum_j \frac{N_j}{k R_j^2} |f_j(\pi)| \exp(-2\sigma_j^2 k^2) \exp(-2R_j/\lambda) \sin(2kR_j + \psi_j + 2\delta)$$

Here N_j is the number of atoms (all of the same type) in shell j with backscattering factor $f_j(\pi)$ at a distance R_j from the central atom. The other terms in Equation 1 are a Debye-Waller like factor σ_j expressing the mean square variation in R_j, the phase factors δ and ψ_j of the photoelectron wave which depend on the central and scattering atom, and λ the mean free path of the photoelectron. The Fourier transform of $k.\chi(k)$ with respect to $sin(2kR)$ or $exp(-2ikR)$ yields a partial radial distribution function in real space with peak areas proportional to N_j. If the phase factors are known, either from theoretical calculations or model compounds (i.e. fitted from the EXAFS of chemically similar compounds to that under investigation but with known R_j and N_j), then the radial distances can be determined, typically out to 5Å from the target atom. The uncertainty in R_j that can be achieved with EXAFS is about +/- 0.01 Å. The determination of N_j is usually less accurate, about +/- 20%, as it is strongly coupled to the Debye-Waller factor. $f_j(\pi)$ does not vary strongly with atomic number and the identification of the type of atoms in the shells is limited to differentiation between rows of the Periodic Table. The advantages of EXAFS over diffraction methods are that it does not depend on long range order, hence it can be used to study local environments in both crystalline and amorphous solids, and liquids, it is atom specific and can be sensitive to low concentrations of the target atom.

II.2. Experimental Procedures

The set up of a typical XAFS experiment is shown schematically in Figure 4. The requirement of a high intensity, tunable X-ray source means that the majority of experiments are now undertaken on synchrotron radiation sources. The experiment is basically simple. The white beam from the synchrotron is passed through a two crystal (usually silicon) monochromator, the wavelength being selected by the Bragg condition and step-wise rotation of the crystals allows a sweep of the X-ray energy. The intensity of the X-ray

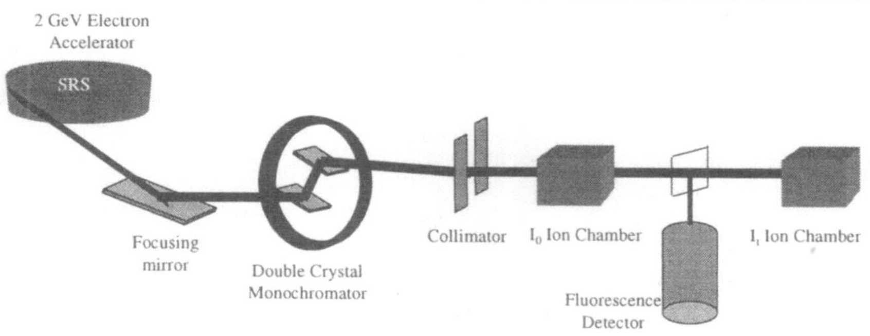

Figure 4. The schematic layout of a typical EXAFS station (9.2 at the Daresbury SRS)

beam incident on the sample, I_o, is measured with an ion chamber filled with a gas mixture set to be 80% transmitting. For samples in which the target atom is concentrated (>1%) a transmission mode is employed and the transmitted intensity, I_t, is measured with a second ion chamber, in this case set to be 80% absorbing. The absorption coefficient is simply evaluated from $log\ (I_o/I_t)$. The sample thickness for a transmission experiment is adjusted to give a step jump at the absorption edge of μ about 1, by using thin films or making pellets of powders with a non-adsorbing diluent (i.e. containing only light atoms, for example boron nitride, silica or polythene). For dilute samples a fluorescence mode is used, where the fluorescence X-rays which are emitted upon absorption of the incident X-rays are used to monitor the absorption coefficient. A scan will typically take the order of 30 minutes. XAFS spectra can be collected for most elements in the Periodic Table, however for the lighter elements (roughly those lighter than calcium) air absorption will significant and the sample needs to be in a vacuum along with the X-ray beam.

The data that will be discussed in later Sections was all collected at the Daresbury SRS which is an electron storage ring operating at 2 GeV and an average current of 150 mA. EXAFS stations 7.1, 8.1, 9.2 or 9.3 were used for the work and these have ion chambers for transmission experiments and multi-element solid state detectors (Canberra) for fluorescence experiments [18]. There are order-sorting monochromators on these stations which allow higher harmonics to be rejected from the incident beam. These stations also allow data to be collected in 'quick-scanning mode' which reduces the scan time to the order of a few minutes. Station 9.3 is a combined station allowing the simultaneous collection of XAS spectra and X-ray powder diffraction (XRPD) patterns from the sample. This station is also equipped with a specially designed furnace so that data can be collected at any temperature between ambient and 1100°C [19]. This has proved particularly useful in the study of the effects of grain growth in nanocrystalline samples.

EXAFS data analysis is performed with interactive computer programmes, such as the EXCALIB, EXBACK and EXCURVE codes developed at the Daresbury SRS [20] or the University of Washington codes, UWAXFS [21]. The work in this chapter was analysed with the Daresbury suite of programmes. EXCALIB allows for correction of the monitors for background counts and converts monochromator angle to X-ray energy. EXBACK removes background X-ray absorption of the atom to produce the normalised absorption, $\chi(k)$. EXCURVE, The final step in the procedure is least-squares fitting the data to a model of the local structure with parameters such as N_j, R_j and A_j [= $2\sigma_j^2$] as variables. Typical graphical outputs from EXCURVE are shown in Figures 5 and 6 which are for the Zr K-edge EXAFS of bulk ZrO_2. Figure 5 is a plot of the experimental normalised absorption coefficient, $\chi(k)$, (depicted by a solid line) as a function of the X-ray wave vector, k, in Å^{-1}, and the value predicted by the theoretical model (depicted by the dashed line). Since the magnitude of the EXAFS oscillations decreases with increasing k it is usual to weight $\chi(k)$ by multiplying by k, k^2 or k^3 to emphasise the oscillations at high k. The iterative least-squares fitting in EXCURVE provides the 'best-fit' to the normalised absorption plot, the quality of the fit being measured by the sum of the deviations between the experiment and the model, usually expressed as a percentage. Figure 6 is a plot of the

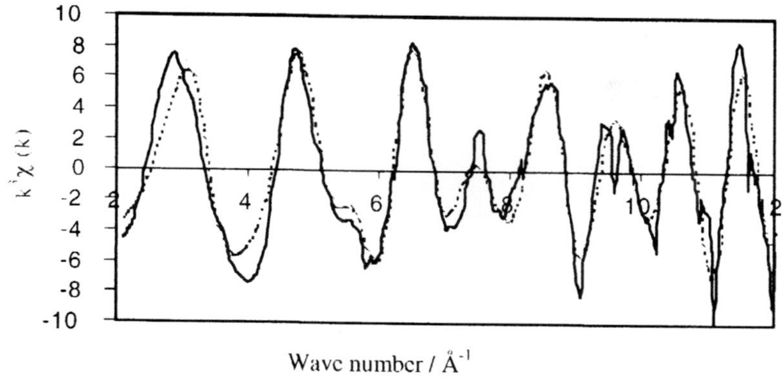

Figure 5. Zr K-edge EXAFS spectrum (k^3 - weighted) for bulk monoclinic zirconia. The solid line is the experimental data and the dashed line the best fit.

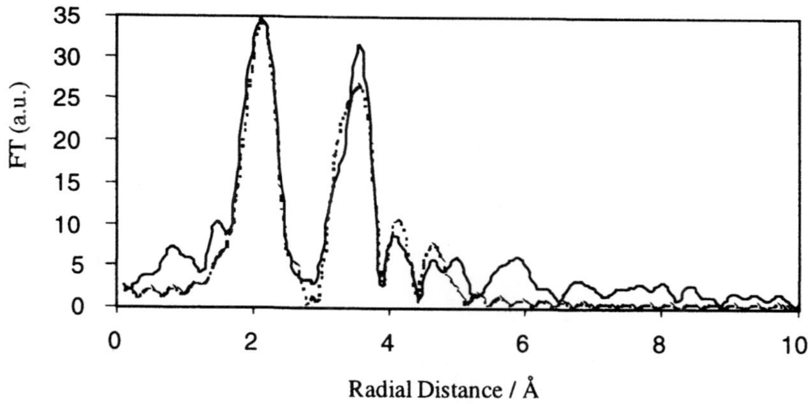

Figure 6. The Fourier transform of the data shown in Figure 5. The solid line is the experimental data and the dashed line the best fit.

magnitude of the Fourier transform as a function of radial distance, *R*. It is important to note that the Fourier transformation has to be performed with a phase-shift, usually that for the atom in the first co-ordination shell. Thus for compounds it can be misleading to read off peak positions from these plots as exact radial distances. Tabulated output from EXCURVE contains the best-fit parameters, the errors on the parameters, the correlation matrix for the parameters and the quality of the fit. EXCURVE has the facility to isolate a peak in the Fourier transform and back transform to yield the normalised absorption, referred to as 'Fourier filtering'. This can be useful in the early stages of an analysis to determine the types of atoms in a co-ordination shell.

III. EXAFS STUDIES OF NANOCRYSTALLINE BINARY OXIDES

III.1. General considerations

Before we consider the individual systems it is worthwhile considering what we might expect the EXAFS spectrum of a nanocrystalline system. The first effect will be that of the grain size affecting (reducing) the average co-ordination numbers of the shells of neighbours and hence attenuating the intensity of the oscillations in the normalised EXAFS spectrum and reducing the size of the peaks in the Fourier transform plot. From the size and shape of the nanocrystals the effect can be calculated from simple geometry. Clearly, the farther the distance of the shell from the target atom the bigger will be the reduction in the average co-ordination number, CN. For the EXAFS of the metal atom in a binary oxide, MO, no effect is expected for the first shell of O atoms. Oxide surfaces in air will be covered by a layer of adsorbed O or OH. The most obvious effect will be on the second shell, the M-M correlation. Figure 7 shows the calculated average value for the second shell CN as a function of grain diameter for some simple oxides with fcc M sub-lattices assuming a cubic geometry for the grains. In view of the EXAFS experimental error on CN (+/-20%) the grain diameter needs to be less than ~3 nm for the effect to be noticeable.

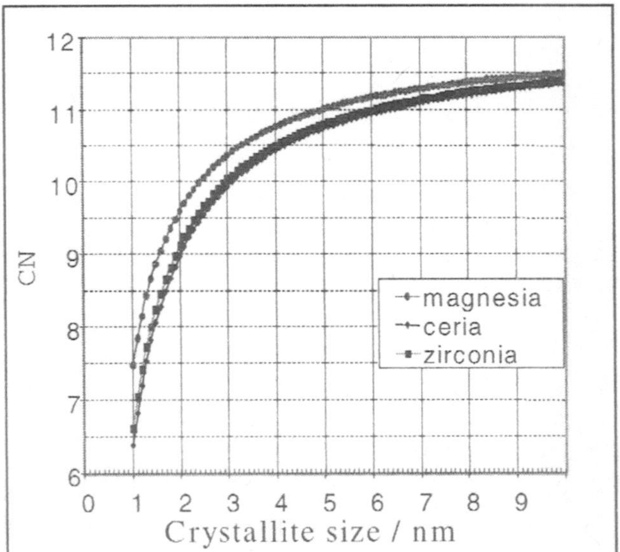

Figure 7. The average cation-cation co-ordination number in nanocrystalline oxides with varying crystallite size.

Disorder in the nanocrystals due to unusually high levels of point defects and/or disorder at the surfaces would also attenuate the EXAFS and reduce the size of the peaks in the plots of the Fourier transform. However, in contrast to the simple grain size effect discussed in the last paragraph, a reduction in the average co-ordination number and an increase in the Debye-Waller factor would attenuate the peaks, the two causes being difficult to deconvulute. High levels of static disorder would be expected to give rise to large Debye-Waller factors, which would be independent of temperature.

From the above discussion the effects that we wish to examine are most easily visualised in the plots of the Fourier transform. Therefore in the following Sections we will mainly present the data in this format, with the Fourier transform corrected with the phase shift of the first shell (oxygen). However, it should be noted that the quantitative data present in the tables has been derived from fitting to the normalised EXAFS spectra using the EXCURVE programme.

Finally two points need to be emphasised concerning the analysis of the EXAFS. Firstly, the average grain size will play an important role in the discussion of the EXAFS. We, along with other workers, have used the peak width of the XRPD patterns, to estimate the grain size on the basis of the Scherrer equation [22]. Secondly, a initial structural model is required for the EXAFS fitting. For this we have used the crystallographic data for the bulk material. We have also collected the EXAFS data for the bulk material and tested our procedures by comparing with the crystallographic data.

III.2. Tin Oxide

Tin oxide occurs naturally as cassiterite and has the rutile structure (space group P42/mnm). It is an *n*-type semiconductor, which is used in flammable gas sensors [23-25]. Strategies that have been employed to improve the sensitivity and selectivity have included the use of nanocrystalline materials and doping with aliovalent cations.

III.2.1. Pure Tin Oxide Powder

Nanocrystalline tin oxide can be prepared by several routes and the method used to prepare the EXAFS samples involved precipitation from an aqueous solution of tin chloride pentahydrate by the addition of aqueous ammonia [26,27]. The (110), (101) and (211) peaks of the XRPD were very broad and the Scherrer equation yields an average particle size of 2 to 3 nm confirming the nanocrystalline nature of the material. The same size was deduced from the reflections of different (*hkl*) values suggesting the particles were of a regular shape with similar dimensions in each direction. Sn K-edge EXAFS spectra were collected in transmission mode at room temperature on station 9.3 at the Daresbury SRS [27].

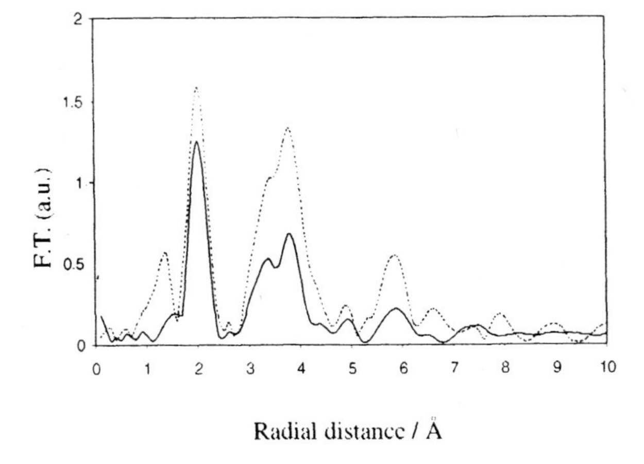

Figure 8. The Fourier transform of the Sn K-edge EXAFS of SnO2; solid line - nanocrystals, dashed line – bulk.

The Fourier transform of the EXAFS data is shown in Figure 8, along with data for bulk SnO_2. Comparison of the two plots shows a dramatic reduction in the magnitude of the second peak, the Sn-Sn correlation. The question arises as to the origin of this reduction.

We took two approaches to the analysis [27] and the results are listed in Table 1. Firstly we analysed the data using the co-ordination numbers for the bulk material. The fit was reasonable, as seen in the R-factor (the goodness of fit) given in the table. The second approach was to float the co-ordination numbers in the fitting to yield the parameter shown in parentheses in Table 1. This improved the R-factor and reduced both the co-ordination numbers and the Debye-Waller factors, the latter becoming similar to those in the bulk material. These co-ordination numbers were consistent with the particle size [29,30] and therefore it was concluded that the reduction of the peaks in the Fourier transform was fully accounted for by this effect alone and there was no requirement to invoke high levels of disorder in the nanocrystals or large numbers of atoms in the interfacial regions.

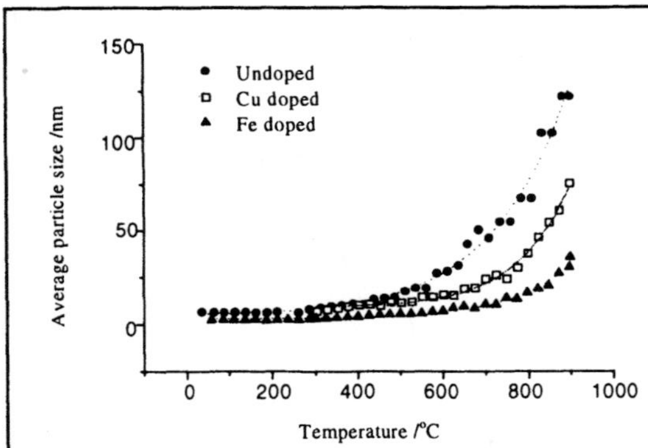

Figure 9. Particle sizes of SnO_2 on heating

There have been a number of EXAFS studies of nanocrystalline SnO_2 [29-31] and the papers are all consistent in explaining the results simply in terms of the particle size. Thus for this system it clear that the boundaries between the nanocrystalline grains are similar to those in bulk material.

The effect of heating on the particle size of the nanocrystals was monitored in parallel with the EXAFS study [27]. Samples were subjected to fixed rate heating on the combined EXAFS/XRPD station. Typical heating rates were 5°C per minute and an XRPD pattern was collected over 2 minutes at 5-minute intervals. The experiments yielded data rapidly and showed that significant growth of the particles began at about 400°C, which is consistent with previous work [30]. Figure 9 shows the particle sizes as a function of temperature for doped and undoped tin oxide collected in synchrotron experiments using the same heating conditions. As expected the growth accelerates with increasing temperature.

Table 1. EXAFS results for pure tin oxide

Atom	*X-ray*[+]		*EXAFS of bulk*			*EXAFS of nano*[*]		
	CN	R/Å	CN	R/Å	A/Å^2	CN	R/Å	A/Å^2
O	2	2.045	6	2.050	0.005	6 *(5.8)*	2.045 *(2.045)*	0.009 *(0.008)*
O	4	2.088						
Sn	2	3.185	2	3.201	0.007	2 *(1.0)*	3.210 *(3.206)*	0.009 *(0.005)*
O	4	3.594	4	3.571	0.001	4 *(2.8)*	3.586 *(3.429)*	0.009 *(0.041)*
Sn	8	3.708	8	3.730	0.007	8 *(4.0)*	3.718 *(3.716)*	0.017 *(0.010)*
			R= 18.44%			R = 17.40% *(14.92%)*		

+ From reference [28]; * The data in parentheses are obtained by floating the co-ordination number.

III.2.2. Doped Tin Oxide Powder

The preparation of nanocrystalline tin oxide can be readily adapted to produce materials doped with cation impurities and EXAFS studies have been made of copper and iron doped materials containing nominally 1, 5 and 10 mole per cent dopant [27].

Cu K-edge EXAFS were collected for Cu doped SnO_2 in fluorescence mode on station 9.3 at the Daresbury SRS. The results were independent of the dopant concentration and those for the nominally 5 mole per cent doped sample will be discussed here. The Fourier transforms for samples following 1 hour annealing at various temperatures are shown in Figure 10. The as-prepared samples at 25°C shows two dominant peaks at ~2 Å and ~4 Å, corresponding to Cu-O and Cu-Sn correlations, respectively. The results of a detailed fitting of the data are given in Table 2. From these parameters and a comparison with EXAFS of copper oxides it is quite clear that in the nanocrystalline SnO_2 samples that the Cu atoms are dissolved predominantly by substitution on the Sn sites [27]. The Debye-Waller factors for the higher shells are higher than those for the pure SnO_2 and would be expected due to the presence of the charge-compensating O^{2-} vacancies.

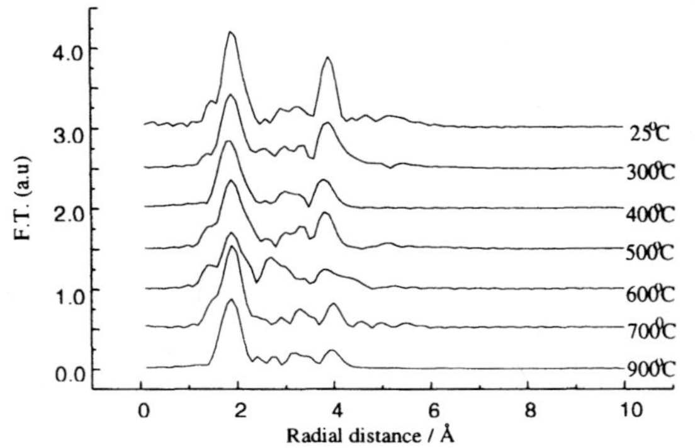

Figure 10. Fourier transforms of nominally 5 mol per cent Cu-doped SnO_2 heated for 1h at high temperatures. Plots have been displaced by 0.5 on the y-axis

Heating the samples causes a reduction in the amplitude of the second peak. There could be a number of explanations of this effect [27]. It is not due to the formation of copper oxide. The simplest explanation, although speculative, is that the grain growth accompanying the annealing results in the Cu atoms are concentrating in the surface regions of the grains, but not necessarily the outermost layer.

Table 2. EXAFS results for doped nanocrystalline tin oxide

Atom	*X-ray for SnO_2*		*Cu K-edge EXAFS Cu-SnO_2*			*Fe K-edge EXAFS of Fe-SnO_2*		
	CN	R/Å	CN	R/Å	A/Å^2	CN	R/Å	A/Å^2
O	2	2.045	6	1.995	0.017	6	2.014	0.008
O	4	2.088						
Sn	2	3.185	2	3.150	0.013	2	3.144	0.011
O	4	3.594	4	3.618	0.011	4	3.460	0.050
Sn	8	3.708	8	3.637	0.018	8	3.688	0.014
			R= 25.68%			R = 35.30%		

The data for the 5 mole per cent Fe doped SnO_2 showed similar features to the Cu doped sample and the interpretation was similar [27]; namely the Fe substitutes for Sn atoms and annealing the samples caused a concentration of the Fe atoms in the surface regions. However, the annealing a 10 mole per cent Fe doped sample at 900°C produced an EXAFS spectrum which was consistent with the precipitation of Fe_3O_4 and the samples visually darkened. This is further evidence for the dopants re-locating in the surface regions.

Finally it is worth noting that the doped samples grew at a slower rate than pure SnO_2 on annealing. This is shown in Figure 8 and has been noted by other workers [32].

III.3. Zinc Oxide

Zinc oxide has the wurzite structure (space group P63mc). Like tin oxide, is an *n*-type semiconductor and has been used in gas sensors [33]. It also has applications as a varistor [34] and in electro-optic devices [35].

III.3.1. Pure Zinc Oxide Powder

Nanocrystalline ZnO can be prepared by reacting the hydroxy chloride (Zn[OH]Cl) with an amine [36,37]. A Reitveld analysis of the diffraction data of this material and TEM data shows the particles are plate-like with dimensions of ~5 nm in the <100> direction and ~3 nm in the <002> direction.

Zn K-edge EXAFS data were collected in transmission mode for bulk and nanocrystalline ZnO on station 7.1 at the Daresbury SRS [37]. In many respects the qualitative nature of the results were similar to those for pure SnO_2 described in Section III.2.1. The most significant feature was an attenuation of the EXAFS oscillations and a marked reduction of the amplitude of the second peak in the Fourier transform, the Zn-Zn correlation. The data for the nanocrystalline sample were analysed by firstly fitting to the crystallographic data with the co-ordination numbers fixed at those in the bulk material. In the second fitting the co-ordination numbers were allowed to float. The results are listed in Table 3. The overall picture is very similar to that for SnO_2. The fitting was improved when the co-ordination numbers were allowed to vary and the final values of these parameters were consistent with the particle size. In addition, the Debye-Waller factors were reduced to values similar to those in bulk ZnO. The CN for the Zn-Zn shell from this fitting is 7.7, compared to 12 in the bulk sample, and it was concluded in the original paper [37] that this was due to a combination of the effects of both particle size and disordered atoms in the surface regions. The value of this CN calculated from the nanocrystal geometry was about 10. In view of the later work on SnO_2, described Section III.2.1, and ZrO_2 and CeO_2 to be discussed in later Sections, together with expected errors in CN of +/- 20% it would now appear reasonable to assign the major contribution of the reduced Zn-Zn CN to the particle size.

Table 3. EXAFS results for pure zinc oxide

Atom	*X-ray*⁺		*EXAFS of bulk*			*EXAFS of nano**		
	CN	R/Å	CN	R/Å	A/Å^2	CN	R/Å	A/Å^2
O	3	1.975	4	1.962	0.008	4 *(3.6)*	1.992 *(1.992)*	0.014 *(0.014)*
O	1	1.988						
Zn	6	3.209	12	3.235	0.016	12 *(7.7)*	3.260 *(3.265)*	0.033 *(0.025)*
O	1	3.219						
Zn	8	3.250						
			R= 32%			R = 46% *(39%)*		

+ From reference [38]; * The data in parentheses are obtained by floating the co-ordination number.

III.3.2. Doped Zinc Oxide Powder

The preparation of nanocrystalline ZnO is readily adapted to yield cation doped material [39-41] and EXAFS studies have been made for Ni doped ZnO and Cu doped ZnO.

The Zn K-edge EXAFS of the samples were similar to that of nanocrystalline pure ZnO discussed in the previous Section. For the Ni K-edge EXAFS of the Ni-doped sample there was a second shell in the Fourier transform, the Ni-Zn (or Ni) correlation, at the distance expected for substitutional solution on the Zn site, but it was considerably reduced in magnitude [40]. A fit indicated there were only 4 cations in the shell compared to 12 in bulk ZnO. In contrast, the Cu K-edge EXAFS for the Cu doped sample showed no second shell [41]. The most reasonable explanation of these data are that the dopants, particularly in the case of Cu, are predominantly located in the surface regions of the crystallites.

III.4. Zirconium Oxide

Zirconium oxide, zirconia, occurs naturally as the mineral baddeleyite which has a monoclinic structure, *m*, (space group P121/c1). The phase diagram of zirconia is complex. At room temperature the stable phase is monoclinic but on heating it transforms at 1400K to the tetragonal form, *t*, (space group P42/nmc) and then at 2640K to a cubic, fluorite-structured phase, *c*, (space group Fm3-m). The addition of lower valent cationic dopants at low concentrations will stabilise the tetragonal phase and at concentration exceeding ~8 mol per cent the cubic phase is stabilised. Although the monoclinic phase is the stable room temperature phase when the material is prepared as nanocrystals (> about 30 nm particle size) other phases can form, most commonly the tetragonal form [42]. Zirconia is second only to diamond in terms of hardness and is used as a specialised ceramic [43]. The cubic stabilised material, normally by the addition of yttrium (so-called YSZ) has a high oxygen ion conductivity at ~1000°C and is used in gas sensors [44], and as an electrolyte in solid oxide fuel cells (SOFC) [45] and oxygen separators [46].

III.4.1. Pure Zirconium Oxide Powder

The samples used in the EXAFS studies were prepared by the commonly used route of precipitating the hydroxide from an aqueous solution of a zirconium salt (in this case $ZrOCl_2.8H_2O$) by the addition of ammonia and the subsequent calcination [47,48]. The temperature and the time of the calcination can be used to determine the particle size of the zirconia [49], for example calcining at 500°C produces particles of 10 nm. Before we consider the EXAFS data we have collected for these samples [50-52] some comments are necessary on the method of preparation. The hydroxide precipitate, often referred to as $Zr(OH)_4$, is an amorphous material containing a hydroxylated zirconyl cluster whose precise composition depends on the preparation conditions. Studies of the structural evolution of the zirconia from the hydroxide show it is a complex process [48,52]. The most revealing has been the recent work [52] which used ^{17}O and ^{1}H nuclear magnetic resonance spectroscopy and this showed unless calcination temperatures greater than 500°C were required for complete conversion of the hydroxide to zirconia. Otherwise, the samples contained amorphous and hydroxylated materials, a very significant point in the discussion of the EXAFS results.

We shall concentrate here on the Zr K-edge EXAFS data collected in transmission mode on station 9.3 at the Daresbury SRS [51] and note that the evolution of the EXAFS on transforming from hydroxide to zirconia is detailed elsewhere [52]. EXAFS data were also collected for bulk zirconia

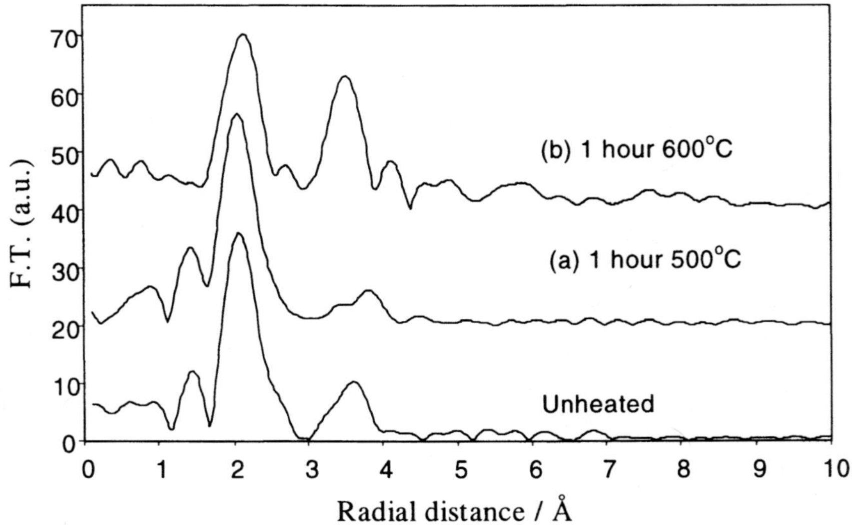

Figure 11. The Fourier transforms of the Zr K-edge EXAFS for the nanocrystalline ZrO_2 powders produced by calcination of zirconium hydroxide. The y-axis has been shifted by 20 units for each successive plot to enhance clarity. The grain size is listed for each plot.

samples, the Fourier transform for the bulk monoclinic sample is shown in Figure 6 and the best-fit parameters for bulk polymorphs are listed in Table 4. These are in excellent agreement with the very thorough EXAFS study of zirconia reported by Li *et al* [53-55] and diffraction data. The Fourier transforms of the hydroxide precursor, and samples calcined for 1 hour at 500°C and 600°C are shown in Figure 11. XRPD data indicated that the 500°C calcined sample contained *t*-phase zirconia particles of ~10 nm diameter and the 600°C contained about 90% *t*-phase and 10% *m*-phase material with a particle size of ~20 nm.

The striking feature of Figure 11 is the very small amplitude of the second peak, the Zr-Zr correlation for the sample calcined at 500°C. X-ray crystallographic data [56,57] for *t* zirconia shows this shell should contain 12 Zr atoms and this would not be significantly reduced for a 10 nm size crystal (see Figure 7). Fitting with the data with a shell of 12 Zr atoms yields a very large Debye-Waller factor of $A = 0.079$ Å^2. It is tempting to assign this value to the presence of a large fraction of atoms in highly disordered interfacial regions, however the true explanation is now known to be the presence of amorphous regions in the sample due to the incomplete conversion of the hydroxide. These data emphasise the need for the careful characterisation of the samples, particularly when the precursor is amorphous and will not be detected in an XRPD pattern.

Table 4 EXAFS results for bulk zirconia standards

Sample /phase		*Zr K-edge*				*Y K-edge*			
		Atom	CN	R/Å	A/Å^2	Atom	CN	R/Å	A/Å^2
ZrO_2	m	O	7	2.15	0.020				
		Zr	7	3.46	0.015				
		Zr	4	4.03	0.017				
		Zr	1	4.55	0.002				
ZrO_2*	m	O	7	2.16	0.020				
		Zr	7	3.46	0.007				
		Zr	4	4.02	0.012				
		Zr	1	4.55	0.005				
$ZrO_2$$	t	O	4	2.10	0.0068	O	8	2.33	0.0154
		O	4	2.33	0.018				
		Zr/Y	12	3.62	0.0068	Zr/Y	12	3.62	0.0072
YSZ#	c	O	7	2.13	0.022	O	8	2.31	0.018
		Zr/Y	12	3.58	0.023	Zr/Y	12	3.61	0.020
YSZ+	c	O	7	2.15	0.019	O	8	2.33	0.018
		Zr/Y	12	3.58	0.022	Zr/Y	12	3.63	0.016

* Data from reference [53]; # 16 atomic % Y; + Data from reference [54] for 20 atomic % Y; $Data from reference [54] for 6 atom % Y at 10K.

III.4.2. Pure Zirconium Oxide Films

These samples, dense films with a thickness of 0.5 μm, were prepared at the University of Missouri-Rolla using a polymer spin-coating procedure on single crystal sapphire substrates. The preparation and characterisation of these films has been described elsewhere [58-62].

Briefly, the procedure is to spin a solution of $ZrOCl_2.8H_2O$ in water/ethylene glycol/glycine on to the substrate, dry the film and calcine to form zirconia. The microstructure depends on the calcining temperature and the grain size can be varied from 1 to 330 nm using temperatures from 300 to 1500°C. The dispersion in the particle size is low, with 80% of the sample being at the average size.

Fluorescence Zr K-edge EXAFS were collected at room temperature on station 9.2 at the Daresbury SRS for pure zirconia films [63,64] for which the grain sizes and structures were determined by XRPD; 5 and 7 nm (*c*), 10 and 22 nm (*t*) and 100 nm (*m*). It should be noted that the presence of the sapphire substrate precluded the use of transmission experiments, however the self-absorption correction was negligible (~1%) [65]. The possibility of cubic and orthorhombic forms of nanocrystalline zirconia has been reported from EXAFS studies [66] and earlier diffraction work [67]. We stress that the primary identification of the phases of our films was based on XRPD. The Fourier transforms of the EXAFS for the samples are shown in Figure 12 and the best-fit parameters are listed in Table 5. The Fourier transforms show no significant reduction of the second peak, the Zr-Zr correlation, and the parameters, even for the smallest grain sizes, are very similar to those of the bulk counterparts. We therefore conclude that the nanocrystals are not highly disordered and that the interfacial regions are comparable to grain boundaries in bulk material. This contrasts with a previous study of the EXAFS of nanocrystalline ZrO_2 and we will defer a discussion of the differences to the next Section.

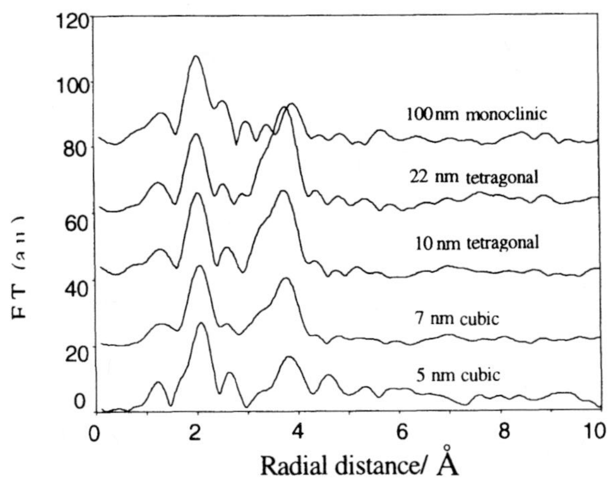

Figure 12. The Fourier transforms of the Zr K-edge EXAFS for the pure ZrO_2 films. The y-axis has been shifted by 20 units for each successive plot to enhance clarity. The grain size is listed for each plot.

Table 5. EXAFS results for pure ZrO_2 films*

Grain size/nm	*Phase*	*Atom*	*CN*	*R/Å*	*A/Å²*
5	*c*	O	8	2.11 (2.13)	0.031 (0.022)
		Zr	12	3.60 (3.58)	0.029 (0.023)
7	c	O	8	2.13 (2.13)	0.034 (0.022)
		Zr	12	3.62 (3.58)	0.024 (0.023)
10	*t*	O	4	2.09 (2.11)	0.012 (0.013)
		O	4	2.31 (2.30)	0.056 (0.037)
		Zr	12	3.62 (3.64)	0.020 (0.020)
22	*t*	O	4	2.10 (2.11)	0.013 (0.013)
		O	4	2.33 (2.30)	0.042 (0.037)
		Zr	12	3.64 (3.64)	0.018 (0.020)
100	*m*	O	7	2.11 (2.15)	0.024 (0.020)
		Zr	7	3.50 (3.46)	0.036 (0.015)
		Zr	4	3.97 (4.03)	0.015 (0.017)
		Zr	1	4.46 (4.55)	0.001 (0.002)

* Values in parentheses are for the bulk counterparts

III.4.2. Yttrium Doped Cubic Zirconium Oxide (YSZ) Films

The spin-coating technique described in the previous Section was modified by adding the appropriate amount of $Y(NO_3)_3$ to produce samples of YSZ containing 16 atom per cent of Y. The grain sizes of the samples were 6, 15 and 240 nm on single crystal sapphire substrates.

Both Zr K-edge and Y K-edge EXAFS were collected in fluorescence mode for the samples at room temperature on station 9.2 of the Daresbury SRS. The Fourier transforms are shown in Figures 13 and 14. The best-fit parameters are listed in Table 6. The results show that there is very little difference between all the samples and there is no significant reduction in the magnitude of of the second peak, the Zr-Zr/Y or Y-Zr/Y correlation, even in the 6 nm sample. There are some differences between the Zr and Y EXAFS results, in particular the second peak is smaller for the Fourier transform of the Zr EXAFS. This has been observed in bulk samples and was interpreted as disorder created by the charge compensating O vacancies being located adjacent to Zr rather than Y ions [68], a view confirmed in later work [54,55].

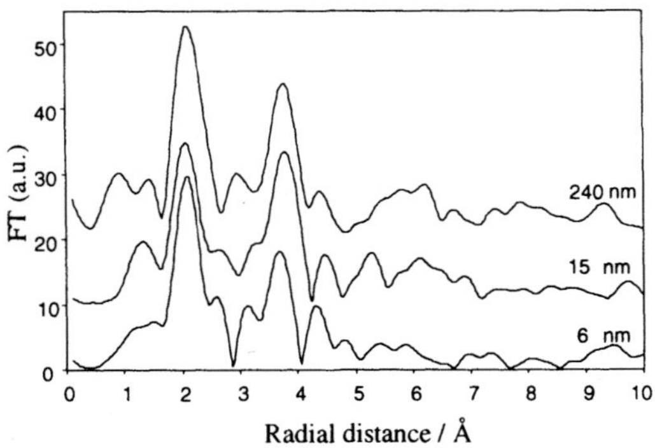

Figure 13. The Fourier transform of Zr K-edge EXAFS of YSZ films. Plots have been displaced by 10 on the *y*-axis.

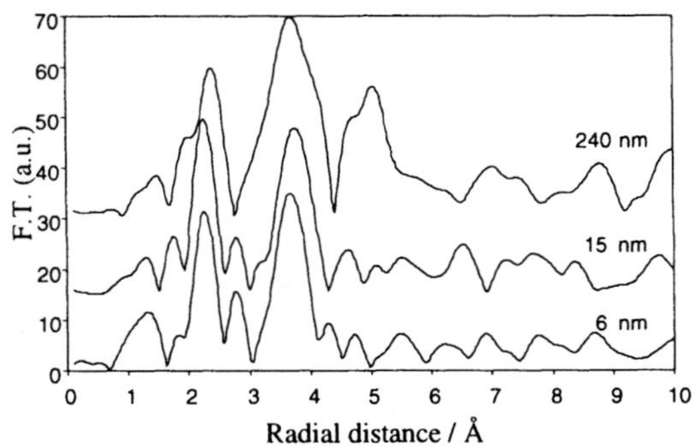

Figure 14. The Fourier transform of Y K-edge EXAFS of YSZ films. Plots have been displaced by 15 on the *y*-axis.

The EXAFS results for the YSZ and those for the pure ZrO_2 films described in the previous section are consistent in showing no major differences between the nanocrystals and the bulk material. There is no significant reduction in magnitude of the second peak in the Fourier transform, as would be expected from the size of the particles; even at 3 nm the Zr-Zr CN would only be reduced from 12 to 10 (see Figure 7) and this would be difficult to detect given a +/-20% error in CN. The conclusion is therefore that the nanocrystals in the films are not highly disordered and that the interfacial regions are similar to grain boundaries in bulk material.

There have been EXAFS studies of ZrO_2, which have shown marked reduction in the Zr-Zr correlation in Fourier transforms. Wang *et al* [66] observed a marked decrease in the magnitude of this peak in decreasing the

Table 6. EXAFS results for YSZ films

Size /nm	*Zr K-edge EXAFS*				*Y K-edge EXAFS*			
	Atom	CN	R/Å	A/Å^2	Atom	CN	R/Å	A/Å^2
6	O	8	2.15	0.020	O	8	2.31	0.020
	Zr/Y	12	3.56	0.029	Zr/Y	12	3.61	0.019
15	O	8	2.17	0.025	O	8	2.30	0.017
	Zr/Y	12	3.61	0.025	Zr/Y	12	3.62	0.020
240	O	8	2.17	0.018	O	8	2.31	0.018
	Zr/Y	12	3.60	0.026	Zr/Y	12	3.61	0.020
bulk	O	8	2.13	0.022	O	8	2.31	0.018
	Zr/Y	12	3.58	0.023	Zr/Y	12	3.61	0.020

particle size from 11 to 5 nm and suggested a possible cause was significant lattice relaxation. The samples were prepared by calcination of co-precipitates, however no details on the temperatures employed were reported in the paper. Therefore we believe the result should be treated with caution given the possible presence of amorphous material due to incomplete conversion to zirconia. The results are similar to those we found for zirconia powders shown in Section III.4.1 and in which we know there is amorphous, hydroxylated material [52]. Deng *et al* [69] reported Zr EXFS data for 14 nm nanocrystalline zirconia and the reduction in the Zr-Zr correlation led to a much reduced CN; 7.2 for loosely compacted powder and 5.9 for compacted powder compared to 12 for the bulk material. This was interpreted as evidence for unusual disorder at the interfaces. However, we would again urge caution in accepting this explanation as no details were given on the materials synthesis. More recently studies of the Zr K-edge of nanocrystalline ZrO_2 [70,71] prepared by evaporation of ZrO noted that the broad asymmetry of the peaks in the Fourier transform of 5 and 8 nm nanocrystals was due to polymorphism. It was concluded that there was no evidence for substantial disorder in the interfaces, in agreement with the results recorded here for films. Finally, we note that Zr K-edge EXAFS studies of ball-milled YSZ, with particle sizes of 10-20 nm, showed a marked reduction in the Zr-Zr correlation although no quantitative analysis was reported [72]. HRTEM micrographs showed nearly perfect crystalline regions surrounded by poorly crystalline and amorphous material. The latter material may be a result of the method of preparation and therefore would not contradict the results for the YSZ films.

III.5. Cerium Oxide

Cerium oxide, ceria, has the cubic fluorite-structure (space group Fm3-m) when pure or doped with lower valent cations. The doped material can be a better oxygen ion conductor than YSZ at equivalent temperatures (e.g. Gd-doped CeO_2) and offers the potential for an SOFC operating at lower temperatures [73].

III.5.1. Pure Cerium Oxide Powders

Nanocrystalline CeO_2 was prepared in a similar manner to ZrO_2 and followed the procedure described in previous studies [74,75]. Cerium hydroxide was precipitated by adding ammonia to an aqueous solution of cerium sulphate, the pH adjusted to 9.0 and the system left to age for 7 days at room temperature. The precipitate was then washed and dried and calcined to the oxide. Calcining at 150, 400 and 600°C produced CeO_2 with particle sizes of 2.4, 4.5 and 12 nm, respectively, as determined by XRPD.

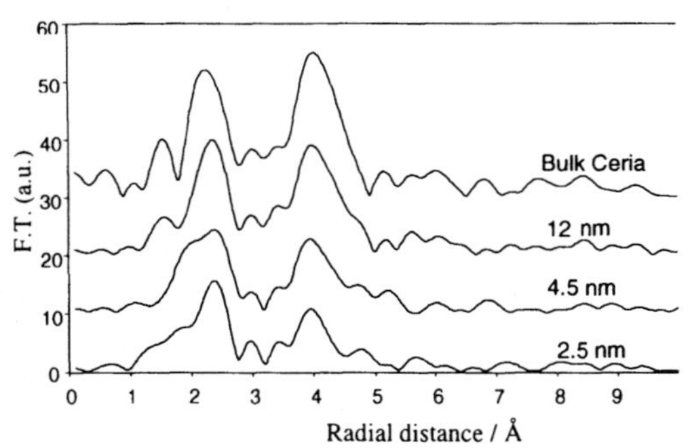

Figure 15. The Fourier transform of Ce L(*III*)-edge EXAFS of CeO_2 powders. Plots have been displaced by 10 on the *y*-axis

Ce *L(III)*-edge EXAFS were collected in transmission mode at room temperature on station 8.1 at the Daresbury SRS. The Fourier transforms of the spectra are shown in Figure 15 and the best-fit parameters in Table 7. In all the plots there is a pronounced second peak, due to the Ce-Ce correlation. In this respect the results are different to those for the ZrO_2 powders discussed in Section III.4.1. There is some reduction in the magnitude of this peak with decreasing particle size and this is reflected in the parameters in Table 7. In deriving the parameters the co-ordination numbers were kept fixed at the crystallographic values, i.e. 8 for the Ce-O shell and 12 for the Ce-Ce shell. Thus the attenuation in the peaks in the Fourier transforms with decreasing size is reflected with increasing Debye-Waller factors. For the 2.5 nm nanocrystals a reduction in the Ce-Ce CN from 12 to ~8 is expected due to the particle size (see Figure 7) and using this value in the fitting reduces the Debye-Waller factor. However, given the known presence of amorphous material in ZrO_2 when it is produced by calcining the hydroxide some doubt must remain about the integrity of the samples. We will return to this point in the next Section.

Table 7. EXAFS results for pure CeO_2 powders

Grain size/nm	*Atom*	*CN*	*R/Å*	*A/Å²*
2.5	O	8	2.32	0.057
	Ce	12	3.83	0.032
4.5	O	8	2.33	0.052
	Ce	12	3.83	0.025
12	O	8	2.33	0.042
	Ce	12	3.84	0.017
Bulk	O	8	2.33	0.030
	Ce	12	3.84	0.15
X-ray Diffraction*	O	8	2.34	
	Ce	12	3.83	

* Data from reference [76]

Nachimuthu *et al* performed a similar Ce *L(III)*-edge EXAFS studied of nanocrystalline CeO_2 prepared by the method outlined above [75]. They noted a reduction in the peaks in the Fourier transform with decreasing size. Their fitted parameters showed a reduced CN for the Ce-Ce correlation for the smallest particle sizes, i.e. 6 and 9 for 2.1 and 4.6 nm particles, which would not be inconsistent with particle size, although this was not mentioned in the original paper. However, the values of the Debye-Waller factor were 3 times higher than for the bulk material and the authors suggested that these smaller particles were less ordered than the bulk. This is a point that we will discuss in the next Section.

III.5.2. Pure Cerium Oxide Films

Nanocrystalline films were prepared on sapphire substrates at the University of Missouri, Rolla by the spin-coating technique use to prepare the ZrO_2 films. The starting material was $Ce(NO_3)_3.6H_2O$ and the particle sizes were 4, 6, 20 and 330 nm.

Ce *L(III)*-edge EXAFS were collected in fluorescence mode at room temperature on station 8.1 at the Daresbury SRS. The Fourier transforms of the spectra are shown in Figure 16 and the best-fit parameters in Table 8.

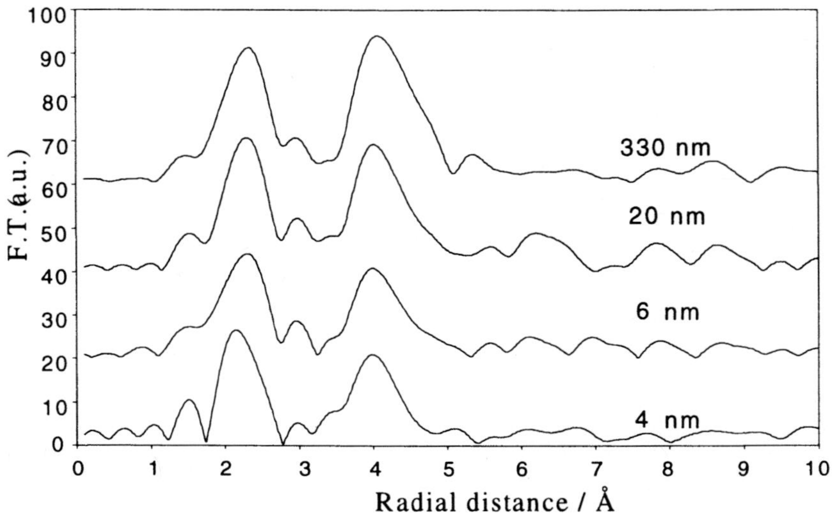

Figure 16. The Fourier transform of Ce L(*III*)-edge EXAFS of CeO_2 films. Plots have been displaced by 20 on the y-axis.

Table 8. EXAFS results for pure CeO_2 films

Grain size/nm	*Atom*	*CN*	*R/Å*	*A/Å²*
4	O	8	2.30	0.027
	Ce	12	3.82	0.020
6	O	8	2.31	0.029
	Ce	12	3.82	0.018
20	O	8	2.32	0.020
	Ce	12	3.83	0.012
330	O	8	2.34	0.019
	Ce	12	3.85	0.009
bulk	O	8	2.34	0.030
	Ce	12	3.84	0.015

For these samples there is no obvious attenuation of the peaks in the Fourier transforms and the best-fit parameters do not vary significantly with particle size. Thus following the arguments presented in Section III.4.2 for the ZrO_2 films we conclude that the CeO_2 nanocrystals are not disordered and that the interfacial regions are similar to grain boundaries in bulk crystals.

It is interesting to compare the results for these films and the nanocrystalline CeO_2 powders, particularly the 4 nm film and the 4.5 nm powder. The Debye-Waller factors for the Ce-O and Ce-Ce shell are higher for the powders, especially for the first of these shells. We have no reason to doubt the integrity of the films and, given the experience with ZrO_2 films and powders, we suspect the possible presence of amorphous hydroxylated material in the smaller nanocrystalline CeO_2 powders. The calcination temperature used to produce the 2.5 nm films is only 150°C, which may be insufficient to completely remove hydroxyl species. Until a thorough study of the preparation has been conducted, in the manner undertaken for ZrO_2 [52], it would be foolhardy to assume the EXAFS of the CeO_2 powders as evidence for disorder in the smaller nanocrystals.

III.5.3. Gadolinium Doped Cerium Oxide Films

Nanocrystalline films of 20 atom per cent Gd-doped CeO_2 were prepared on sapphire substrates at the University of Missouri, Rolla by the spin-coating technique use to prepare the CeO_2 films. The starting materials were $Ce(NO_3)_3.6H_2O$ and $Gd(NO_3)_3.6H_2O$ the particle sizes were 4, 9, and 15 nm.

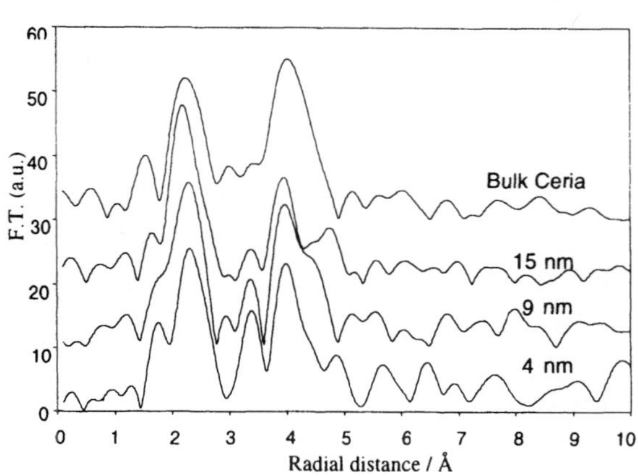

Figure 17. The Fourier transform of Gd L(*III*)-edge EXAFS of Gd-doped CeO_2 films. Plots have been displaced by 10 on the *y*-axis. The Ce L(*III*)-edge data for bulk CeO_2 is shown for comparison.

Gd *L(III)*-edge EXAFS were collected in fluorescence mode at room temperature on station 8.1 at the

Daresbury SRS. The Fourier transforms of the spectra are shown in Figure 17 and the best-fit parameters in Table 9. All the spectra were very similar and there was no obvious dependence on crystallite size on the magnitude of the second peak in the Fourier transform. This is reflected in the parameters shown in Table 9. The Gd-O distance is slightly larger than the Ce-O distance in pure CeO_2, 2.36 Å compared to 2.30 Å, in films of similar grain size. This is probably an effect of ionic radii, Gd^{3+} radius is 1.053 Å and Ce^{4+} radius is 0.97 Å. In contrast, the Gd-Ce distance, 3.79 Å, is slightly less the Ce-Ce distance, 3.83 Å in pure CeO_2, possibly due to the difference in cation charge. However, at this relatively high level of doping the Gd ions will be present as clusters. The structures of the possible clusters have been discussed in recent EXAFS studies of doped CeO_2 [77,78].

Table 9. Gd *L(III)*-edge EXAFS results for Gd doped CeO_2 films

Grain size/nm	*Atom*	*CN*	*R/Å*	*A/Å^2*
4	O	8	2.36	0.025
	Ce/Gd	12	3.79	0.016
9	O	8	2.36	0.025
	Ce/Gd	12	3.79	0.020
15	O	8	2.36	0.026
	Ce/Gd	12	3.79	0.026

IV. CONCLUSIONS AND OVERVIEW

This Chapter has described EXAFS studies of pure and doped samples of simple binary oxides with grain sizes as small as 3 to 4 nm. In this Section we will summarise and examine this work on the basis of three objectives. The first objective is a clear presentation of the major conclusions of the experiments. The second objective is to examine the connection between physico-chemical properties and microstructure in the light of the EXAFS work. The final objective is to look at areas and experiments where there could be potential for future developments. Clearly this will be personal view of the authors and speculative, but we feel it is a necessary inclusion in this Section.

The first point we wish to emphasise is the requirement for care in both the preparation and characterisation of oxides prior to EXAFS studies. This is especially important when oxides are prepared by the calcination of hydroxides, as demonstrated by the work on ZrO_2 [51,52], since complete conversion to the oxide requires higher temperatures than are often assumed. EXAFS monitors the environment of the *average* target atom and the presence of unreacted precursor or reaction intermediates can mask the spectrum of the true oxide. We believe that some of the discrepancies amongst the various EXAFS studies of nanocrystalline oxides can be rationalised on this basis.

The major conclusion we draw from the EXAFS work in this Chapter is that in all the four oxides investigated the attenuation of the cation EXAFS spectrum of nanocrystals can be explained simply on the basis of the reduced CN due to the small particle size. There is no requirement to invoke high levels of disorder in the nanocrystals or a large fraction of atoms in the interfacial regions. Therefore we conclude that boundaries between the nanocrystals are similar to grain boundaries in normal bulk material. Although some of the early EXAFS studies of nanocrystalline metals claimed evidence of highly disordered interfaces a very careful study of nanocrystalline Cu [79] also concluded the interfaces were like normal grain boundaries. In the case of the oxides this result is not surprising, as the strong Coulomb forces would militate against disorder even in surface and intergrain regions. For example, studies of the surface 'rumpling' of oxides, distortion of the bond lengths in the surface, suggest the effect is very small [80,81] with bond lengths at the surface being perturbed by ~0.1 Å.

This picture of the microstructure of a nanocrystalline oxide as well-ordered crystallites separated by normal grain boundaries, clearly differs from that presented at the beginning of this article in Figure 1 and must be reconciled with the unusual properties of the materials. A first point to note is that although the nature of the grain boundaries is normal their density is extremely high and this will undoubtedly affect the transport and mechanical properties of nanocrystalline samples. In addition, these boundaries can act as sinks in which considerable amounts of impurity can be finely dispersed leading to apparently high impurity solubilities. Secondly, in the case of ionic materials grain size alone can be responsible for a number of unusual phenomena when the size is comparable or less than the depth of the space-charge layer. This will perturb the point defect equilibrium [10], although not to a level detectable in a conventional EXAFS experiment, as has been beautifully demonstrated in the recent study of the ionic conductivity of thin fluoride films [11]. A third point is that although the crystallites are ordered and their gross crystallographic structures can be determined, there is little information on their surface morphology. Surface morphology can be crucial in determining chemical reactivity and there are indications that it may be very different in nanocrystals than in bulk material. It has been suggested that high index faces are exposed on MgO nanocrystals leading to enhanced reactivity with gases [8,9]. In addition, the nanocrystal surfaces may contain a higher concentration of active sites, steps and edges that can enhance reactivity

[82,83].

The location of the dopants in nanocrystalline oxides depends on the individual system. In cases like Y-doped ZrO_2 and Gd-doped CeO_2 the dopants have a high solubility and there appears to be no difference between the nanocrystalline and bulk samples. However, in both types of sample at high dopant levels there is extensive clustering of the dopants [77,78]. Nanocrystalline Fe and Cu-doped SnO_2, in their as-prepared and unheated states, show substitutional location of the dopants on the Sn site. On heating and subsequent growth of the grains, the dopants become located in the surface regions of the crystallites. The process by which this migration of the dopant occurs is unclear but is worthy of further study as it is important in the application of the materials. The two dopants investigated in nanocrystalline ZnO show differing behaviour; Ni appears to substitute for Zn, whereas Cu is located in the surface regions.

A final experimental observation worthy of note is that the nanocrystalline oxides begin to grow at moderate temperatures, growth being apparent in SnO_2 and ZrO_2 at 400°C. The grain growth can be retarded in SnO_2 by the presence of cation impurity [27]. Grain growth presents a problem in experiments and applications, which require elevates temperatures to produce measurable effects, a good example being the use of nanocrystalline SnO_2 in flammable gas-sensing [84].

Finally, we turn to the areas of potential future development and progress. EXAFS experiments will continue to play an important role in the examination of the microstructures of nanocrystalline oxides. In addition to extending the measurements to other binary oxides a particularly informative study would be comparative EXAFS measurements on nanocrystals of the same oxide produced by different routes (e.g. chemical, vapour deposition, ball-milling, etc.). Such a study should identify microstructural differences, if there are any, between the samples. It is envisaged that more chemically complex oxides will be prepared in nanocrystalline form to explore their potential for applications. EXAFS measurements, with their special ability to probe the environments of each atomic component in the oxide, will be particularly important in the characterisation of these new materials.

In the discussion above the morphology of nanocrystals has emerged as an area where further work should be fruitful. Firstly, there is important work to be undertaken in gaining an understanding of why the morphology of some nanocrystals differs from the bulk parent. The surface energy must be playing a crucial role but the details are not clear. An ideal example is that of pure zirconia where the structure changes from cubic to tetragonal to monoclinic with increasing size. This is the type of problem that is now amenable to computer simulation techniques [85]. These methods are now well advanced, particularly for ionic materials [86], and it is feasible to use energy minimisation methods to model the structure of a complete nanocrystalline oxide, i.e. a particle of 1 nm diameter. The surface morphology of nanocrystals is important in understanding their chemical reactivity and further AFM and STM studies should be informative. This surface morphology will also be revealed in computer simulations and a comparison with the results from the microscopy will provide an important test of the methods.

This Chapter has focused on the use of one particular technique to study a rather simple class of nanocrystalline material, the binary oxides. The understanding of the microstructure of these systems is now in a good state although there are several details that still need to be resolved. An exciting area where future growth is to be expected is in the physical and chemical applications of these materials.

ACKNOWLEDGEMENTS

We wish to acknowledge the various students and co-workers at Kent who have worked on nanocrystalline materials, in particular and in alphabetical order, Helen Brook, Annette Harsch, Kevin Kennedy, Manuel Melle-Franco, David Maitland, Norberto Morgante, Georgios Rafeletos, Nigel Russell, Shelley Savin, Mark Smith, John Strange, Matt Tucker, Alan Wilson, Adam Whitham and John Wright. We are very grateful for help over several years from staff at the Daresbury SRS that has facilitated the work described in this Chapter, especially staff associated with the EXAFS beam lines; Neville Greaves, Andy Dent, Ian Harvey, Bob Bilsborrow, Fred Mosselmans, Lorrie Murphy and Nick Chinnery. We owe special thanks to Igor Kosacki at the University of Missouri, Rolla. He provided the thin film samples of zirconia and ceria that been characterised by XRD in his laboratory (the cubic zirconia films had not been previously reported and we obtained the samples prior to publication). The films were decisive in our gaining an understanding of the nanocrystalline oxides and we are deeply indebted to him not only for the samples but also for endless good-natured help and advice.

REFERENCES

[1] Gleiter, H. *Adv. Mater.*, **1992**, *4*, 474.
[2] Henglein, A., *Chem. Rev.*, **1989**, *89*, 1061.
[3] Weller, H., *Angew. Chem., Int. Ed. Engl.* **1993**, *32*, 41.
[4] Seigel, R.W. and Fougere, G.E., *Nanostructured Mater.* **1995**, *6*, 205.
[5] Gleiter, H., *Acta Mater.*, **2000**, *48*, 1.
[6] Efros, A.L. and Rosen, M., *Ann. Rev. Mat. Sci.*, **2000**, *30*, 475.
[7] see, for example, papers in the special issue of *Chem. Mat.*, **1996**, *8*, no. 8.

[8] Stark, J.V., Park, D.G., Lagadic, I. and Klabunde, K.J., *Chem. Mat.,* **1996**, *8*, 1904.

[9] Stark, J.V. and Klabunde, K.J., *Chem. Mat.,* **1996**, *8*, 1913.

[10] Maier, J., *Prog. Solid State Chem.,* **1995**, *23*, 171.

[11] Sata, N, Eberman, K., Eberl, K. and Maier, J., *Nature*, **2000**, *408*, 946.

[12] Chadwick, A.V., *Nature,* **2000**, *408*, 925.

[13] Teo, B.K. and Joy, D.C., eds. "*EXAFS Spectroscopy; Techniques and Applications*", Plenum Press, New York, **1980**.

[14] Hayes, T.M. and Boyce, J.B., *Solid State Phys.*, **1982**, *37*, 173.

[15] Koningsberger, D.C. and Prins, R., eds. "*X-Ray Absorption*", Wiley, New York, **1988**.

[16] Catlow, C.R.A. and Greaves, G.N., eds. "*Applications of Synchrotron Radiation*", Blackie, Glasgow, **1990**.

[17] Lee, P.A. and Pendry, J.B., *Phys. Rev. B*, **1975**, *11*, 2795.

[18] Details of the stations can be found on the Daresbury web site; URL http:\\www.dl.ac.uk

[19] Dent, A.J., Oversluizen, M., Greaves, G.N., Roberts, M.A., Sankar, G., Catlow, C.R.A. and Thomas, J.M., *Physica*, **1995**, *209*, 253.

[20] Binsted, N., Campbell, J.W., Gurman, S.J. and Stephenson, P.C., *SERC Daresbury Program Library*, Daresbury Laboratory, Warrington, Cheshire WA4 4AD, UK, 1992; Dent, A.J. and Mosselmans, J.F.W., *Guides to EXCALIB, EXBACK and EXCURV92*, Darebury Laboratory, Warrington, Cheshire WA4 4AD, UK, 1995.

[21] Stern, E.A., Newville, M., Ravel, B., Yacoby Y. and Haskel, D., *Physica B*, **1995**, *209*, 117.

[22] Klug, H.P. and Alexander, L.E., "*X-Ray Diffraction Procedures*", Wiley; New York, **1974**.

[23] Lantto, V., in "*Gas Sensors; Principles, Operations and Developments*", ed. Sberveglieri, G.; Kluwer; Dordrecht, **1992**, p.117.

[24] Williams, D.E., in "*Solid State Gas Sensors*", ed. Moseley, P.T. and Tofield, B.C.; Adam Hilger; Bristol; **1987**, p71.

[25] Taguchi, N., *UK Patent Specification*, 1280809, **1970**.

[26] Ansari, S.G., Boroojerdian, P., Kulkarni, S.K., Sainkar, S.R., Karekar, R.N. and Aiyer, R.C., *J. Mat. Sci. - Materials in electronics*, **1996**, *7*, 267.

[27] Davis, S.R., Chadwick, A.V. and Wright, J.D., *J. Phys. Chem .B.*, **1997** *101*, 9901.

[28] Seki, H., Ishizawa, N., Mizutani, N. and Kato, M., *Yogyo Kyokai Shi*, **1984**, 92, 219.

[29] Briois, V., Santilli, C.V., Pulcinelli, S.H. and Brito, G.E.S., *J. Non-Cryst. Solids*, **1995**, *191*, 17.

[30] Brito, G.E.S., Briois, V., Puncinelli, S.H. and Santilli, C.V., *J. Sol-Gel Sci. and Tech.*, **1997**, *8*, 269.

[31] Brito, G.E.S., Ribeiro, S.J.L., Briois, V., Dexpert-Ghys, J., Santilli, C.V. and Puncinelli, S.H., *J. Sol-Gel Sci. and Tech.*, **1997**, *8*, 261.

[32] Xu, C., Tamaki, J., Miura, N. and Yamazoe, N., *Chem. Lett.,* **1990**, p. 441.

[33] Heiland, G., *Sensors and Actuators,* **1982**, 2, 343; Morrison, S.R., *Sensors and Actuators,* **1982**, *2*, 329.

[34] Einzinger, R., *Ann. Rev. Mat. Sci.,* **1987**, *17*, 299.

[35] Look, D.C., *Mat. Sci. Eng B.,* **2001**, *80*, 383; Mais, N., Reithmaier, J.P., Forchel, A., Kohls, M, Spanhel, L. and Muller, G. *Appl. Phys. Lett.*, **2001**, *75*, 2005.

[36] Garcia-Martinez, O., Rojas, R.M., Vila, E. and Martin de la Vidal, J.L., *Solid State*

Ionics, **1993**, *63-65*, 442.

[37] Chadwick, A.V., Russell, N.V., Whitham, A.R. and Wilson, A., *Sensors and Actuators B*, **1994**, *18*, 99.

[38] Kisi, E.H. and Elcombe, M.M., *Acta Cryst. C*, **1989**, *45*, 1867.

[39] Whitham, A.R., *M.Sc. thesis*, University of Kent, **1993**.

[40] N.V. Russell, A.V. Chadwick and A. Wilson, *Nucl. Inst. & Methods B*, **1995**, *97*, 575.

[41] Chadwick, A.V., Harsch, A., Russell, N.V., Tse, K.F., Whitham, A.R. and Wilson, A., *Rad. Eff. Def. Solids*, **1995**, *137*, 1277.

[42] Garvie, R.C., *J. Phys. Chem*, **1965**, *69*, 1238; Garvie, R.C., *J. Phys. Chem*, **1978**, *82*, 218.

[43] Stevens, R, "*An Introduction to Zirconia*", Magnesium Elektron Ltd, London, **1986**.

[44] Steele, B.C.H., Drennan, J., Slotwinski, R.K., Bonanos, N. and Butler, E.P., in *Advances in Ceramics*, eds. A.H. Heuer, and L.W. Hobbs, American Ceramic Society; Columbus OH, **1981**; Vol 3, p. 286.

[45] Badwal, S.P., *Appl. Phys.*, **1990**, *50*, 449.

[46] Ziehfreund, A. and Maier, W.F., *Chem. Mat.*, **1996**, *8*, 2721.; Steele, B.C.H., Comptes Rendu, C-Chimie, **1998**, *1*, 533.

[47] Clearfield, A., *Rev. Pure Appl. Chem.*, **1964**, *14*, 91.

[48] Turrilas, X., Barnes, P., Gascoigne, D., Turner, J.Z., Jones, S.L., Norman, C.J., Pygall, C.G. and Dent, A.J., *J. Radiat. Phys. Chem.*, **1995**, *45*, 491.

[49] Liu, H., Feng, L., Zhang, X. and Xue, Q., *J. Phys. Chem.*, **1995**, *99*, 332.

[50] Brook, H.C., Chadwick, A.V., Kennedy, K.M., Morgante, N., Rafeletos, G., Tomba, A. and Roberts, M.A., *Material Science Forum*, **1997**, *239-241*, 687.

[51] Rush, G.E., Chadwick, A.V., Kosacki, I. and Anderson, H.U., *J. Phys. Chem. B.*, **2000**, *104*, 9597.

[52] Chadwick, A.V., Mountjoy, G., Nield, V.M., Poplett, I.J.F., Smith, M.E., Strange, J.H. and Tucker, M.G., *Chem. Mater.*, **2001**, *13*, 1219.

[53] Li., P., Chen, I-W, and Penner-hahn, J.E., *Phys. Rev. B*, **1993**, *48*, 10063.

[54] Li., P., Chen, I-W, and Penner-hahn, J.E., *Phys. Rev. B*, **1993**, *48*, 10074.

[55] Li., P., Chen, I-W, and Penner-hahn, J.E., *Phys. Rev. B*, **1993**, *48*, 10082.

[56] Teufer, G., *Acta Cryst.*, **1962**, *15*, 1187.

[57] Michel, D, Mazerolles, L. and Jorba, M.P.Y., *J. Mater. Sci.*, **1983**, *18*, 2618.

[58] Chen, C.C., Nasrallah, M.M. and Anderson, H.U., U.S. Patent 3,494,700, February, **1996**.

[59] Chen, C.C., Nasrallah, M.M. and Anderson, H.U., *Solid State Ionics*, **1994**, *70/71*, 101.

[60] Kosacki, I., Gorman, B., and Anderson, H.U., *Proc. Third Int. Symp. Ionic and Mixed Conducting Ceramics*, ed. T.A. Ramanarayanan, *Electrochem. Soc. Proc.*, **1997**, *97-24*, 631.

[61] Kosacki, I., Petrovsky, V. and Anderson, H.U., *Appl. Phys. Lett.*, **1999**, *74*, 341.

[62] Kosacki, I, Shumsky, M. and Anderson, H.U., *Ceram. Eng. Soc. Proc.*, **1999**, *3*, 135.

[63] Rush, G.E., Ph.D. thesis, University of Kent, **2001**.

[64] Rush, G.E., Chadwick, A.V., Kosacki, I. and Anderson, H.U., *in preparation.*

[65] Dent, A.J., *private communication.*

[66] Wang, Y.R., Lu, K.Q, Wang, D.H., Wu, Z.H. and Fang, Z.Z., *J. Phys.; Condensed Matter*, **1994**, *6*, 633.

[67] Srinisan, R., de Angelis, R.J., Ice, G. and Davis, B.H., *J. Mater. Res.*, **1991**, *6*, 1287.

[68] Catlow, C.R.A., Chadwick, A.V., Greaves, G.N. and Moroney, L.M., *J. Am. Ceram. Soc.*, **1986**, *69*, 272.

[69] Deng, H., Qiu, H. and Shi, G., *Physica B*, **1995**, *208&209*, 591.

[70] Nitsche, R., Winterer, M., Croft, M. and Hahn, H., *Nucl. Inst. & Methods B*, **1995**, *97*, 127.

[71] Winterer, M., Nitsche, R. and Hahn, H., *Nanostructured Mater.*, **1997**, *9*, 397.

[72] Michel, D., Gaffet, E. and Berthet, P., *Nanostructured Mater.*, **1995**, *6*, 647.

[73] Kharton, V.V., Figueiredo, F.M., Navarro, L., Naumovich, E.N., Kovalevsky, A.V., Yaremchenko, A.A., Viskup, A.P., Carneiro, A., Marques, F.M.B. and Frade, J.R., *J. Mat. Sci.*, **2001**, *36*, 1105.

[74] Hiano, M. and Kato, E., *J. Mater. Sci. Lett.*, **1996**, *15*, 1249.

[75] Nachimuthu, P., Shih, W.C., Liu, R.S., Jang, L.Y. and Chen, J.M., *J. Solid State Chem.*, **2000**, *149*, 408.

[76] Wolcyrz, M. and Kepinski, L., *J.Solid State Chem.*, **1992**, *99*, 409.

[77] Yamazaki, S., Matsui, T., Ohashi, T. and Arita, Y., *Solid State Ionics*, **2000**, *136-137*, 913.

[78] Ohashi, T., Yamazaki, S., Tokunaga, T., Arita, Y., Matsui, T., Harami, T. and Kobayashi, K., *Solid State Ionics*, **1998**, *113-115*, 559.

[79] Stern, E.A., Siegel, R.W., Newville, M., Sanders, P.G.. and Haskel, D., *Phys. Rev. Lett.*, **1995**, *75*, 3874.

[80] Goniakowski, J. and Noguera, C., *Surface Sci.*, **1995**, *323*, 129.

[81] Goniakowski, J., Holender, J.M., Kantorovich, L.N., Gillan, M.J. and White, J.A., *Phys. Rev. B.*, **1996**, *53*, 957.

[82] Koper, O.B., Lagadic, I., Volodin, A. and Klabunde, K.J., *Chem. Mater.*, **1997**, *9*, 2468.

[83] Carnes, C.L. and Klabunde, K.J., *Langmuir*, **2000**, *16*, 3764.

[84] S.R. Davis, A.V. Chadwick, and J.D. Wright, *J. Mater. Chem.*, **1998**, *8*, 2065.

[85] Catlow, C.R.A. and Jackson, R.A., *private communication.*

[86] Catlow, C.R.A., ed., "*Computer Modeling in Inorganic Crystallography*", Academic Press, New York, **1997**.

SURFACE ANALYSIS OF SEMICONDUCTING NANOPARTICLES BY FTIR SPECTROSCOPY

Application to the Evaluation of Gas Sensing Properties

Marie-Isabelle BARATON
SPCTS – UMR 6638 CNRS, University of Limoges, Limoges (France); baraton@unilim.fr

1. INTRODUCTION

From the 1960s, the development of new characterization techniques has made possible the investigation of surfaces at molecular and atomic levels (1). However, most of these techniques only allow the analysis of clean surfaces of crystals under high vacuum. If the crystal has to be used in an environment other than vacuum, the structure and the chemical composition of the crystal surface will necessarily evolve to balance forces at the interface between the crystal and its surrounding milieu. Adsorption and eventually absorption of various molecules from the ambient result in the introduction of foreign atoms at the very surface and possible relaxation and reconstruction processes can modify the crystalline structure of the surface layer. As a consequence, the study of clean surfaces is not always sufficient for a complete understanding of the surface chemistry under the real working conditions of the materials. The control of adsorption, absorption and reconstruction phenomena is not an easy task because they depend on a large number of parameters such as chemical composition of the environment, temperature and pressure, crystalline structure of the exposed planes, impurities possibly present on the surface as residues of the synthesis conditions.

Nanosized particles (or nanoparticles) can be considered as surfaces in three dimensions (1) and, therefore, a thorough characterization of their surface becomes an essential prerequisite toward controlled, reproducible and optimized properties. The strong curvature of nanoparticle surfaces makes the standard representation of the surface by the terrace-step model no longer applicable. The terraces which are ordered domains are reduced to a very small size, thus greatly increasing the concentration of kinks and corners where the atoms are in low coordination number and thus highly reactive. As a consequence and independently from the fact that the relative number of surface atoms is larger, the surface of nanoparticles is more reactive than that of their micron-sized counterpart. This constitutes both a tremendous advantage and an important drawback. Indeed, in fields involving surface phenomena such as catalysis and gas sensing, the use of nanoparticles should bring improvements in the properties due to the larger exposed surface and the higher surface reactivity. In return, this high surface reactivity leads to easy and fast surface contamination by the ambient and the contaminating species may have adverse effects on the surface properties. This is the reason why precise characterization and control of surfaces are absolutely essential for high-added-value applications of nanoparticles. On the other hand, it can be taken advantage of the high surface reactivity of nanoparticles to engineer their surface composition and chemistry by grafting selected organic molecules, thus tailoring the surface properties for a defined application. Obviously, any surface modification process needs careful monitoring.

An important criterion in the choice of the surface characterization technique is the depth of sampling. The depth of sampling depends on several factors such as the energy of the probing particles (electrons, ions, photons), the angle of the incident beam, the nature of the material to be analyzed. Moreover, when nanoparticles have to be analyzed, this depth of sampling should be significantly smaller than the particle size to solely take the surface layer into account. On the other hand, the thickness of the "useful" surface layer is determined by the considered phenomenon (2). For example, adsorption only concerns the adsorbed species and the first atomic layer of the materials whereas the thickness variation of the depletion layer in a semiconducting material usually concerns several hundreds of atomic layers in depth. So, to investigate a surface phenomenon, the choice of the surface technique must be made by carefully comparing the thickness of the surface layer perturbed by the considered phenomenon with the depth that the envisaged technique can sample. Whereas the adequacy can be fairly easily obtained for the investigation of a single phenomenon, the choice of the characterization technique becomes tremendously complex when several phenomena affecting the surface at different thicknesses have to be simultaneously studied.

In addition, if the nanoparticles have to be used in atmosphere or in an environment other than vacuum, their surface must be analyzed under conditions as close as possible to the working conditions. The choice of techniques suitable for surface characterization of nanoparticles in that quasi-real environment is rather limited and Fourier transform infrared (FTIR) spectroscopy is among these few techniques. Although FTIR spectroscopy is not usually considered as a surface technique, it can nevertheless provide very useful information on the surface chemical composition and reactivity of nanoparticles under various atmospheres and at different temperatures provided a specific setup is attached to the spectrometer. In addition, *in situ* experiments can be performed, thus allowing a monitoring of surface chemical modifications. This way of using FTIR spectroscopy is quite familiar to scientists working in catalysis. But, in addition, we have proved that FTIR spectroscopy can also be extremely useful for the study of semiconducting nanoparticles by giving information on the variations of the electrical conductivity of the sample when its gaseous environment is varied.

In this chapter, after a brief overview of the fundamentals of surface FTIR spectroscopy, it will be described how this technique allows the study of the chemical reactions at a semiconductor surface simultaneously with the consequences on the electrical conductivity changes. As examples, results on titanium oxide and tin oxide will be presented. These two materials are semiconductors and are widely used in the fabrication of chemical gas sensors.

2. FUNDAMENTALS OF THE CHARACTERIZATION TECHNIQUE

It is well-known that the infrared spectrum of a molecule originates from the absorption of the infrared energy at discrete wavelengths, thus exciting the interatomic vibrations. In the harmonic oscillator approximation, the intensity of the absorption bands depends on the reduced mass of the vibrating atoms and on the force constant of the bonds. On the other hand, according to the Beer's law, the absorbance is directly related to the concentration of the absorbers. Details on infrared spectroscopy, including Fourier transform spectroscopy and spectrometers can be found in (3-6).

When an infrared beam crosses a particle, the transmitted beam contains information not only on the interatomic bonds constituting the bulk, but also on the chemical groups at the particle surface. Obviously, the latter are minority by far. But, when the size of the particle is sufficiently decreased, down to the nanometer scale for example, the concentration of the surface groups relatively to that of the bulk interatomic bonds significantly increases and the contribution of these surface groups to the overall infrared absorption becomes no longer negligible. This is the reason why FTIR spectrometry can be conveniently used for the surface analysis of nanoparticles. However, in most cases, the absorption bands due to the surface groups are muck weaker than those due to the bulk modes. To further increase the absorption of the surface groups, their concentration has to be increased. This is achieved by analyzing the powdered sample without any dilution in a salt such as potassium bromide. But under these conditions, the absorptions from both bulk and surface groups are simultaneously enhanced, obviously resulting in a noticeable absorption of the surface groups whereas the sample becomes opaque in the wavenumber region where the bulk modes absorb. Fortunately, the surface groups are usually organic-like species with several vibrational modes absorbing above 1000 cm^{-1} whereas the interatomic vibrations in the bulk of a ceramic material usually absorb below 1200 cm^{-1}. As a consequence, simultaneous analysis of bulk and surface is hardly possible. In any case, precise experiments must be performed to discriminate the absorption bands due to surface species from those originating from the bulk (overtones/ combinations of bulk modes and eventually impurities trapped in the bulk or close pores) (7).

2.1. FTIR Surface Characterization

The detailed experimental procedure for surface characterization of nanosized particles can be found elsewhere (8-13). It is briefly summarized in the following.

The sampling technique consists in slightly pressing the pure nanosized powder into a thin self-supported pellet. Depending on the nature of the material and its specific surface area, the amount of powder needed varies from 20 to 100 mg. The pressure applied to obtain a suitable pellet is adjusted for each material and must cause no change either in the crystalline structure or in the stoichiometry of the original nanopowder. It is worth noting that the pressure to be applied may vary for a material depending on its synthesis conditions and on the environment because surface groups are dependent on these parameters. A stainless grid can conveniently be used for supporting the pellet, additionally allowing a homogeneous temperature distribution if the sample needs to be heated.

The pellet is placed in the small furnace of a vacuum cell. This cell specially designed to fit in the sample chamber of the spectrometer allows the IR spectrum to be recorded *in situ* at any temperature (from room temperature to 773 K) under vacuum or gaseous atmospheres. The cell is connected to vacuum pumps through a nitrogen trap and to gas cylinders and liquid containers. Defined pressures of gases or liquid vapors can be directly introduced inside the cell by using a precision valve system while selected mixtures of gases (or liquid vapors) can be prepared in a separated compartment before introduction.

The first step of the surface characterization is a heat-treatment under dynamic vacuum (referred to as *activation*) to remove the surface species that are physisorbed and weakly chemisorbed. These adsorbed species result from the environmental contamination and balance the forces at the interface of the material and its environment. Once these adsorbed species are eliminated, the surface is no longer in equilibrium and becomes *activated*. In other words, an activated surface, which can only exist under vacuum, presents freed adsorption sites ready to adsorb any new molecules. The nature of the adsorption sites on an activated surface depends on the activation conditions (temperature and vacuum). Therefore, the activation conditions must be clearly described and taken into account when comparing the adsorption and reactive sites on the same material characterized by different researchers. In addition, it must be clearly understood that an *activated* surface is different from a *clean* surface. Indeed, a clean surface has never been exposed to environmental contamination (e.g. synthesis or crystal cleavage under ultra-high vacuum conditions) whereas an activated surface is only partly freed from contaminating species as the activation treatment does not usually remove all the foreign atoms introduced by the environmental contamination. As a basic example of irreversible contamination, we can quote nitrides or carbides surfaces which are partly oxidized after exposure to regular atmosphere and which usually consist in oxinitride or oxicarbide layers, respectively (8, 14-19). An attempt at complete cleaning by an increase of the activation temperature may result in a surface reconstruction and eventually grain growth. When semiconducting nanoparticles are to be activated, additional consequences of the heat-treatment process are observed on the IR spectrum and will be discussed in the following section (cf. section 2.2).

The remaining chemical groups on the activated surface must be discriminated from possible impurities trapped in the bulk. The most probable contaminating surface groups are hydroxyl (OH) groups due to the dissociation of atmospheric water. Also, carbonate (CO_3^{2-}), nitrate (NO_3^{2-}) and methyl groups (CH_3) are often found, originating most of the time from synthesis residues and they are partly eliminated by activation. However, specific experiments must be performed to prove that the chemical groups persisting after activation are actually at the nanoparticle surface. Because most of surface groups contain hydrogen, the usual way to check their surface location is to exchange hydrogen by its isotope, deuterium. Due to the higher molecular weight of deuterium, all the vibrations involving hydrogen in the exchanged groups will shift toward lower wavenumbers. Therefore, the H/D isotopic exchange acts as a marker of the surface groups containing hydrogen. Carbonate and nitrate surface groups are usually identified by their characteristic absorption frequencies and by their perturbation following the adsorption of probe-molecules as explained below.

The third step of the surface characterization consists in the controlled adsorption of selected molecules (referred to as *probe-molecules*) on the adsorbing and reactive sites freed by the activation. The vibration frequencies of the adsorbed molecule will be perturbed with respect to the gas phase. These perturbations depend on the nature of the adsorption which can be physical adsorption (physisorption) or chemical adsorption (chemisorption). In the first case, the interaction between the probe-molecule and the surface is of Van der Waals type and the perturbations in the probe-molecule spectrum are relatively weak. But in case of chemisorption, the spectrum of the probe-molecule can be strongly modified particularly if the interaction leads to a surface reaction. In addition, these perturbations in the probe-molecule spectrum also depend on the nature of the surface atoms to which the molecule is bonded and on the coordination number of these atoms within the surface. Therefore, by comparing the IR spectrum of the molecule in the adsorbed form and in the gas phase, information on the adsorbing and reactive sites on the nanoparticle surface can be obtained as a function of the activation temperature. Moreover, the use of a broad range of probe-molecules with different reactivities will allow the full characterization of the surface sites in terms of acidic and basic strength. Examples of molecules that are conveniently used to probe the surface sites can be found in Ref. 7, 12, 20, 21.

2.2. FTIR Analysis of Semiconducting Nanoparticles

The semiconducting property originates from the mobility of free carriers which are electrons (n-type semiconductors) or holes (p-type semiconductors). According to the Drude-Zener theory (22, 23), these free carriers contribute to the absorption by the material over the whole infrared range. This has been experimentally demonstrated by Harrick (24) who applied an electric field to a silicon wafer to modify the space charge region while measuring the absorption of an infrared beam. Harrick observed variations of the infrared energy absorbed by the silicon wafer which were directly related to the variations of the free carriers concentration. In other words, when the free carrier density increased, the absorption of the silicon wafer over the total infrared range increased. The absorption of the free carriers gives rise to an additional characteristic band (intraband) in the infrared spectrum of the semiconductor. Most theories (23-26) predict roughly a λ^2 dependence for this absorption (λ being the infrared wavelength) although variations often occur. But, the general rule is that the background absorption of the infrared radiation by a semiconductor sample increases when the electrical conductivity of the semiconductor increases. At the upper limit, a conductor sample (like a metal) is totally opaque to the IR radiation.

In the case of a chemical gas sensor based on semiconducting materials, the variations of the electrical conductivity are induced by adsorption of oxidizing or reducing gases on the semiconductor surface (27). The adsorption of oxygen, for example, on a semiconductor, causes a decrease of the electron density in the conduction band due to the formation of negatively charged oxygen species (ionosorbed species), such as O^- or O_2^-. The oxygen ionosorption causes electron transfer from the surface of the grain toward the adsorbed species, thus leading to the formation of an electron-depleted surface layer (space-charge region). As a result, the electrical conductivity of a n-type semiconductor decreases due to the enhancement of the potential barrier at the grain boundaries. On the contrary, when a reducing gas, such as CO, adsorbs, electrons are injected into the conduction band and the electrical conductivity of the n-type semiconductor increases. The results of these adsorptions on the infrared spectrum of the n-type semiconductor are a decrease of the overall absorption under oxidizing gases and an increase of the overall absorption under reducing gases. In terms of infrared energy transmitted by the sample, an increase of the electrical conductivity leads to a decrease of the transmitted infrared energy whereas a decrease of the electrical conductivity leads to an increase of the transmitted infrared energy.

Considering that, on the one hand, FTIR spectroscopy allows one to follow *in situ* the reactions taking place at the gas-nanoparticle interface (cf section 2.1), and that, on the other hand, the variations of the background infrared absorption are related to the variations of the electrical conductivity, it becomes clear that FTIR surface spectroscopy is an excellent method to fundamentally study the gas detection mechanisms by semiconductor-based sensors. It must be stressed that, in this particular case, FTIR spectrometry combines the investigation of two related phenomena (surface reactions and electrical conductivity) involving different thicknesses of the material as the surface reactivity concerns only the first atomic layer whereas the thickness variation of the depletion layer may concern up to several tens of nanometers.

The heat-treatment under dynamic vacuum (activation) of semiconducting nanopowders usually leads to oxygen desorption, thus resulting in an increase of the electron density, that is an increase of the free carrier density in a n-type semiconductor material such as tin oxide. As a consequence, a n-type semiconductor material becomes opaque to the infrared radiation under activation. It is therefore easy to understand that the standard activation process is not always appropriate for surface characterization of n-type semiconducting nanopowders. To overcome the difficulty, we have defined a slightly modified activation procedure: the semiconducting nanopowder is heated under dynamic vacuum up to 423 K to eliminate physisorbed water and possible weakly bound impurities. Then, the sample is heated up to the chosen activation temperature under a few mbar of oxygen, thus avoiding oxygen desorption. At the activation temperature, the cell is evacuated to remove the contaminants which have possibly desorbed from the surface during heating and a new dose of fresh oxygen is introduced to stabilize the stoichiometry. The semiconducting nanopowder is then cooled under oxygen. This procedure has been applied to the tin oxide nanopowder which is discussed in following. However, in the case of titanium oxide, the increase of the electrical conductivity during activation is not as critical as in the case of tin oxide and, the titanium oxide nanopowder was activated under the standard conditions previously defined (cf. section 2.1).

3. SURFACE CHARACTERIZATION

Before discussing the gas sensing properties of the titanium oxide and tin oxide nanopowders, a short description of the chemical groups that can be found at the surface is given below.

3.1. Titanium Oxide

The titanium oxide nanopowder (titania, n-TiO_2) whose characteristics and properties are presented in this chapter, is the commercial product P25 from Degussa-France. It is mainly in the anatase crystalline phase (~70%). The specific surface area measured by the supplier is 50±15 m^2g^{-1} with an estimated average size of the primary particles of 21 nm. For the present infrared analyses, 50 mg of n-TiO_2 powder are slightly pressed into thin pellets on a stainless grid (Gantois, France).

The infrared spectrum of the nanopowder at room temperature and under vacuum is presented in Figure 1a in the 4000-1000 cm^{-1} range. At wavenumbers lower than 1200 cm^{-1}, the strong absorption of the bulk modes completely obscure the spectrum.

After activation at 723 K (Figure 1b) under standard conditions (cf. section 2.1), the disappearing of the very intense band centered around 3320 cm^{-1} is observed, indicating the elimination of water molecules bonded to the surface hydroxyl groups. At the end of the activation treatment (Figure 1b), several bands in the 3780-3600 cm^{-1} range are observed. All of them are assigned to ν(OH) stretching vibrations (28-35) of surface OH groups. The multiplicity of these bands indicates that several types of OH groups are present on the titania surface. According to the literature (28-35), the number and the vibrational frequencies of these OH groups depend on the extent of the dehydroxylation and on the possible presence of impurities, which makes the comparison between different samples and between the results from different works extremely difficult. Moreover, the OH groups on the anatase surface are not exactly the same as those on the rutile surface (33). It must also be taken into account in the bands assignment that a hydroxyl group can be bonded to one titanium atom, or linked to two titanium atoms, or bridged to three titanium atoms. As a consequence, all these OH surface groups have a different reactivity and it has been proved that some of them are Brönsted acid sites whereas some other ones are Brönsted base sites (36). Besides, these OH groups are responsible for the surface hydrophilicity (35).

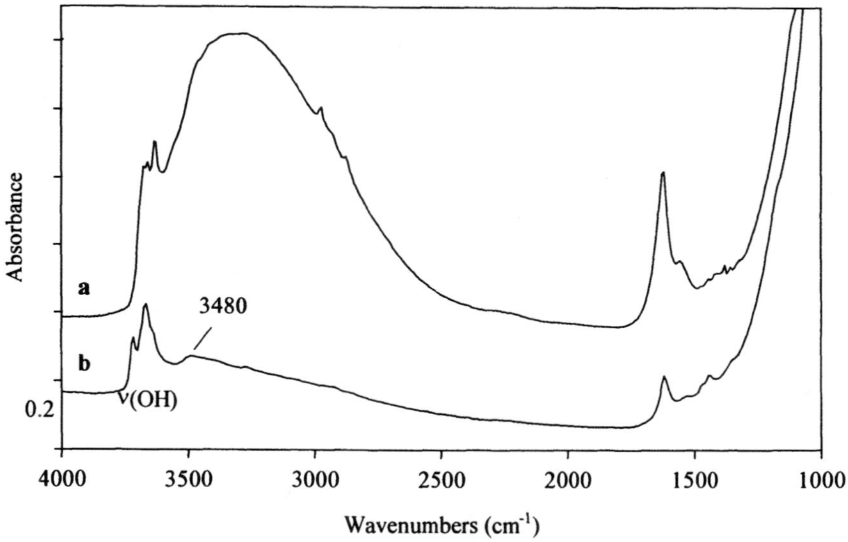

Figure 1. Infrared spectra of the TiO_2 nanopowder: a) at room temperature and under vacuum; b) after activation at 723 K. (The spectra have been shifted for clarity sake).

Despite this complexity, the precise assignment of each absorption frequency to a particular type of surface OH groups can be found in several papers (see for example Ref. 28-35). In addition, the broad band at 3480 cm^{-1} (Figure 1b) may be assigned to adsorbed undissociated water which is retained on the surface by strong Lewis acid sites, namely coordinatively unsaturated Ti^{4+} sites (30, 35). As it will be seen in the following, the surface hydrophilicity may have adverse effects on the reliability of the n-TiO_2 response to gases.

3.2. Tin Oxide

The tin oxide nanopowder (n-SnO_2) investigated in this study has been synthesized by evaporation of compressed micro-crystalline powder with the pulse radiation of a Nd:YAG-laser and subsequent condensation of the vapor in a controlled atmosphere (37). The XRD analysis shows that the nanopowder is crystallized under the quadratic phase with an average particle size of 15 nm. To obtain a sample suitable for FTIR surface analysis, a very light pressure was applied to 30 mg of n-SnO_2 powder on a stainless grid.

As previously described (cf. section 2.2), the tin oxide nanopowder is activated under a low oxygen pressure (50 mbar) to avoid oxygen desorption which would have led to the opacity of the sample to the infrared radiation. Figure 2 compares the spectrum of the n-SnO_2 pellet recorded at room temperature and under dynamic vacuum with the spectrum recorded after activation at 673 K under oxygen. Like in the case of n-TiO_2, a decrease of the broad band centered at 3200 cm^{-1} is observed although weaker. It corresponds to the elimination of water molecules adsorbed on the surface. After activation, we also clearly observe the multiplicity of ν(OH) bands in the 3800-3000 cm^{-1} region, which indicates a large number of different types of surface OH groups (38, 39). The different vibrational frequencies of the OH groups originate from the diversity of the coordination types of the surface tin atoms to which the OH groups are bonded. Indeed, it is known that Sn^{2+} and Sn^{4+} along with oxygen vacancies can be present on the surface (40), thus generating different electronic distributions in the attached OH groups. The bands in the 1500-1000 cm^{-1} region are assigned to the corresponding δ(OH) bending vibrations of these OH groups (41).

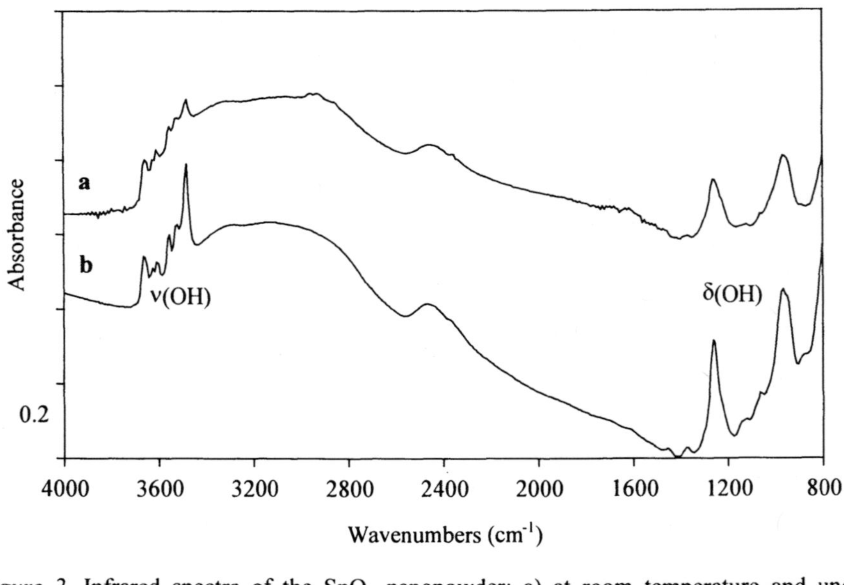

Figure 2. Infrared spectra of the SnO_2 nanopowder: a) at room temperature and under vacuum; b) after activation at 673 K under 50 mbar O_2. (The spectra have been shifted for clarity sake).

4. EVALUATION OF THE GAS SENSING PROPERTIES

As explained in section 2.2, the variation of the infrared energy transmitted by a semiconductor sample when changing its gaseous environment, is related to the variation of the electrical conductivity.

Tin oxide and titanium oxide are n-type semiconductors and among the most popular materials used for the fabrication of chemical gas sensors (42). In chemical sensors based on semiconducting metal oxides, a chemical reaction occurs at the material surface between oxygen species and the gas to be detected, thus modifying the semiconductor resistance (27). Indeed, it is generally admitted that the variations of the semiconductor resistance allowing the detection of combustible gases is caused by a reaction of the combustible gas with ionosorbed oxygen species. The concentrations of the different types of ionosorbed oxygen species, such as O_2^-, O^- and O^{2-}, depend on the temperature. While the O^- species are the most reactive ones in the 573-623 K range (43-45), the O_2^- species become preponderant at lower temperatures (373-473 K). When a combustible gas, such as carbon monoxide, is subsequently adsorbed, oxidation reactions proceed and the electrons trapped by the ionosorbed oxygen species return to the grains, thus decreasing the height of the potential barrier and the material resistance. The chemical reactions leading to changes in the electrical conductivity are usually proposed as follows, depending on the temperature (46-49):

$$CO + O_2^- \rightarrow CO_2 + \tfrac{1}{2}\,O_2 + e^-$$
$$CO + O^- \rightarrow CO_2 + e^-$$
$$CO + O^{2-} \rightarrow CO_2 + 2\,e^-$$

In the following, we study these variations of the electrical conductivity by analyzing the changes in the background infrared absorption of the samples versus gas exposures.

4.1. Tin Oxide

On the activated and oxidized surface of the tin oxide nanopowder (cf. section 2.2), oxygen is first adsorbed at 623 K and then CO is introduced in the cell in presence of oxygen. The comparison of the infrared spectra recorded before and after CO addition (Figure 3a,b) clearly indicates the formation of CO_2 (band centered at 2348 cm^{-1}). In addition, the bands in the 1500 cm^{-1} region are due to newly formed CO_3^{2-} carbonate groups (48-50). These carbonate groups are also observed when, under the same conditions, carbon dioxide is adsorbed on the tin oxide surface.

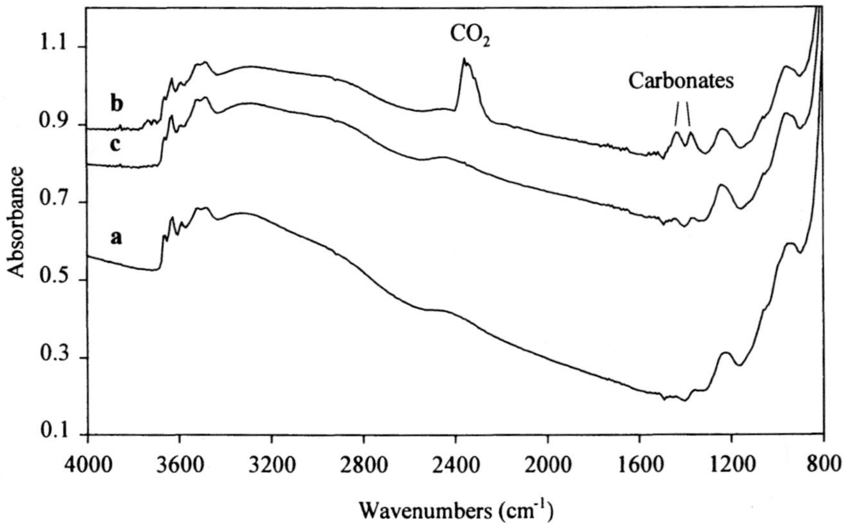

Figure 3. Infrared spectra of the SnO_2 nanopowder recorded at 623 K: a) under O_2 (50 mbar); b) after CO addition (10 mbar); c) after evacuation. (The spectra have NOT been shifted).

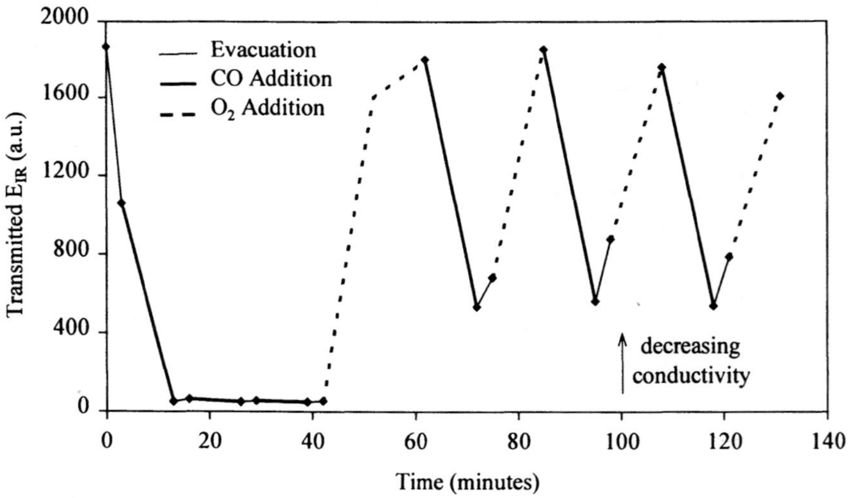

Figure 4. Variations of the infrared energy (E_{IR}) transmitted by the SnO_2 nanopowder at 623 K versus gas exposures.

It can therefore be concluded that CO oxidation into CO_2 proceeds first and then newly formed CO_2 adsorbs on surface basic sites leading to surface carbonate groups, according to the following reaction:

$$CO_2 + O^{2-} \rightarrow CO_3^{2-}$$

It must be noted that, in principle, the formation of carbonate groups does not participate in the variation of the electrical conductivity because the involved electrons remain localized. Only the formation of CO_2 leads to the release of free carriers. Both CO_2 and surface carbonates are eliminated by evacuation (Figure 3c).

In addition to the formation of these new species, an increase of the overall absorption of the sample is observed. Indeed, the baseline of the n-SnO_2 spectrum shifts toward higher absorbance values under CO addition (Figure 3a,b), due to the increase of the free carrier density.

Figure 4 shows the variations of the infrared energy *transmitted* by the tin oxide sample versus gas (O_2 and CO) exposures. This curve can be compared to the sensor response curve (standard impedance measurements) (51). A decrease of the electrical conductivity (corresponding to an increase of the transmitted infrared energy) is observed when oxygen is adsorbed whereas an increase of the electrical conductivity (corresponding to a decrease of the transmitted infrared energy) is caused by CO adsorption. When CO is adsorbed in absence of oxygen, a strong reduction of the SnO_2 sample is observed and the oxidation state is not restored by evacuation. In fact, the sample is so reduced that the addition of a second and third CO dose in absence of oxygen is no longer detected by the material due to a saturation. The reproducibility of the sensor response is ensured by the presence of oxygen allowing a complete recovery of the n-SnO_2 oxidation state after elimination of both CO_2 and carbonate groups by evacuation.

4.2. Titanium Oxide

In this section, we compare the effects of the adsorption of pure CO and of mixtures of CO and water vapor on both the n-TiO_2 infrared spectrum and the evolution of the transmitted IR energy. Because the presence of oxygen as a third gaseous component would have certainly complicated the surface chemical reactions, the experiments were conducted in an oxygen-free environment, taking advantage of the low sensitivity to oxygen of titanium oxide at our selected operating temperature. Therefore, the following discussion should be considered as a preliminary study toward the understanding of the humidity effects rather than an exact description of the real sensor behavior.

In a first step, several doses (6 mbar) of pure CO (referred to as "dry CO") were subsequently adsorbed at 673 K on the n-TiO_2 surface, pretreated at this temperature under the standard conditions described in section 2.1. As expected on this n-type semiconductor, the CO adsorption leads to the sample reduction. This is indicated by the decrease of the transmitted IR energy (Figure 5) due to the increase of the free-carrier density. To emphasize the surface chemical modification possibly occurring under CO adsorption, the difference between the spectra recorded just before CO addition and after 10 minutes under CO is reported in Figure 6. The formation of CO_2 originating from the oxidation of CO and responsible for the sample reduction, is observed. The evacuation of CO only allows a very partial recovery of the oxidation state of the sample because this experiment is performed in absence of oxygen (Figure 5). Subsequent additions of CO doses also lead to a decrease of the transmitted IR energy and, thus to an increase of the electrical conductivity. It is observed that the sample is steadily reduced and that the amplitude of the response toward the CO doses gradually decreases showing the impoverishment of the titania surface in oxygen accessible for CO oxidation. It is worth noting the negative feature on the difference spectrum (Figure 6) at 3650 cm^{-1} indicating that the surface OH groups are perturbed.

In a second experiment still at 673 K, four doses of pure CO (referred to as "dry CO") were adsorbed on the n-TiO_2 activated surface, followed by four doses of a mixture of 6 mbar CO and 1 mbar water vapor (referred to as "wet CO"), then followed by four new doses of "dry CO". (Figures 7 and`8). The response toward the first four "dry CO" doses is obviously similar to that just described (Figures 5 and 7). The addition of "wet CO" doses causes a strong decrease of the electrical conductivity although without any reproducibility. In parallel, the infrared spectra show the intensity increase of the ν(OH) absorption range (Figure 8b). This is actually the proof that water molecules dissociate on the surface:

$$H_2O + O^{2-} \rightarrow 2\ OH^-$$

However, this dissociation reaction may not be the only type of interaction between the surface species/sites and the water molecules. The overall effect of the four "wet CO" doses is a steady oxidation (Figure 7), clearly showing the adverse effect of H_2O. Indeed, the adsorption of "wet CO" doses results in an oxidation of the sample instead of the expected reduction by pure CO. When "dry CO" is added again, the energy evolution is similar to that observed during the addition of the first "dry CO" doses. Moreover, the baseline drift shows the same downward trend, thus indicating an overall reducing effect (Figure 7).

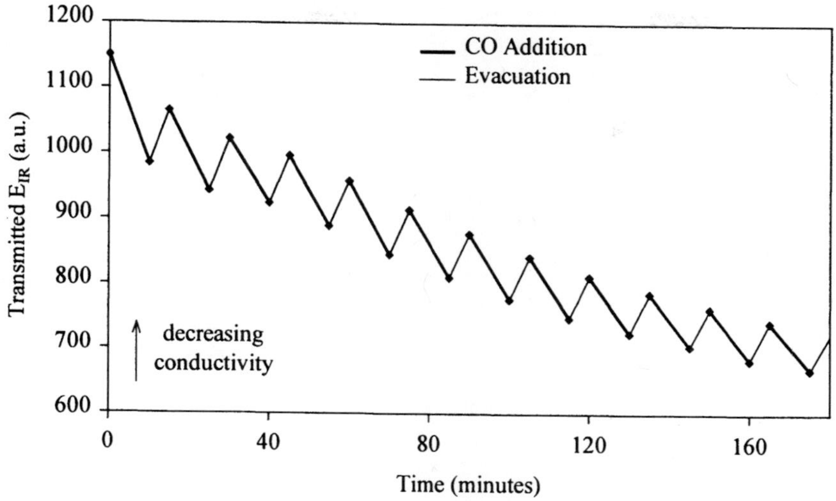

Figure 5. Variations of the infrared energy (E_{IR}) transmitted by the TiO_2 nanopowder at 673 K versus gas exposures.

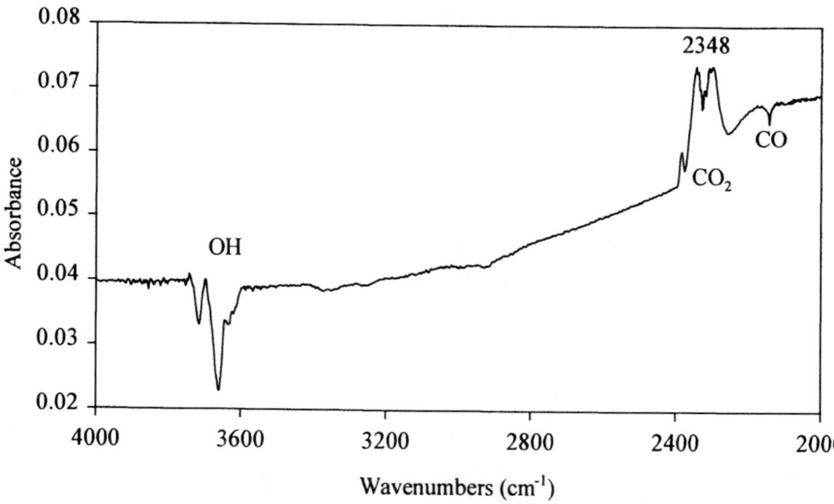

Figure 6. Difference spectrum showing the modifications of the n-TiO_2 surface under CO adsorption at 673 K.

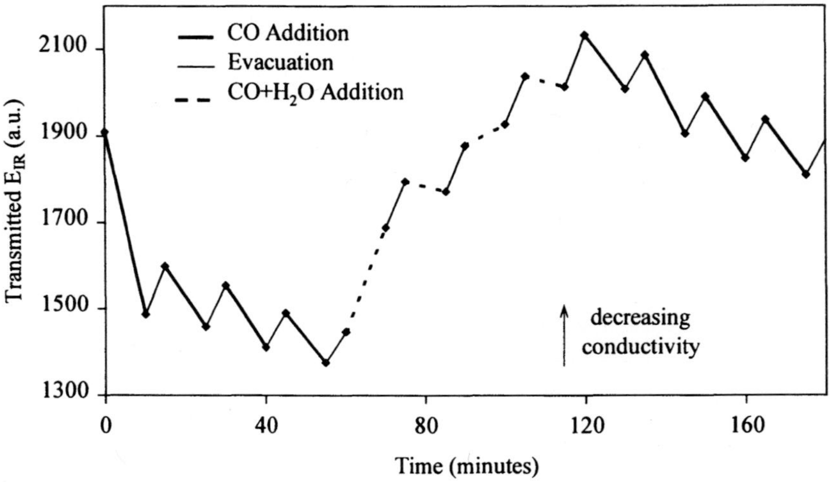

Figure 7. Variations of the infrared energy (E_{IR}) transmitted by the TiO_2 nanopowder at 673 K versus gas exposures.

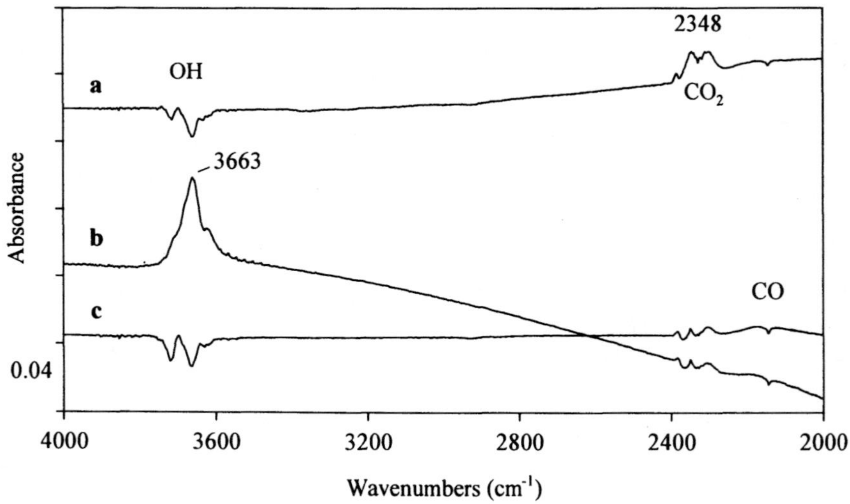

Figure 8. Difference spectra showing the modifications of the n-TiO_2 surface at 673 K: a) under CO adsorption (first series of "dry CO" doses); b) under ($CO+H_2O$) adsorption ("wet CO" doses); c) under CO adsorption (second series of "dry CO" doses). (The difference spectra have been shifted for clarity sake)

5. REDUCTION OF HUMIDITY EFFECTS BY SURFACE MODIFICATION

From the above experiments, it is clear that the presence of humidity is a critical point affecting the reliability of the response of semiconductor-based sensors. The surface OH groups, which should favor an interaction between the surface and the water molecules, do not appear to play a key part, though. To check the influence of the surface hydrophilicity on the sensitivity to humidity, a chemical modification of the surface of the titania nanoparticles has been conducted to make this surface partly hydrophobic. However, on any material, modification of the surface groups affects the surface reactivity and, on semiconductor materials, surface modification can affect the depth of the depletion layer, thus changing the electrical properties (52-54). It must therefore be kept in mind that, for nanoparticles in which the thickness of the depletion layer is comparable to the radius of the nanoparticle, changes in the chemical nature of the surface groups may imply strong consequences on the surface reactivity and on the electrical properties.

5.1. Surface Chemical Modifications

The surface of the titania nanoparticles has been modified by grafting hexamethyldisilazane (HMDS) which is commonly used to make surfaces hydrophobic. The titania nanoparticles were first activated at 673 K. After cooling the sample at room temperature under dynamic vacuum, 7 mbar of HMDS were introduced in the cell for 20 minutes. Then, the sample was steadily desorbed at increasing temperature up to 673 K. This grafting experiment was performed *in situ* so that the chemical modifications of surface species were exactly followed by recording the IR spectrum at each experimental step. Figure 9 compares the n-TiO_2 spectra recorded, at room temperature, after activation (Figure 9a) and after HMDS desorption at 673 K (Figure 9b).

Hexamethyldisilazane usually reacts with surface hydroxyl groups (55-59) by forming ammonia according to the reaction:

$$2\ \text{Ti-O-H} + (CH_3)_3\text{-Si(NH)Si-}(CH_3)_3 \rightarrow 2\ \text{Ti-O-Si-}(CH_3)_3 + NH_3$$

It can indeed be observed on Figure 9b that, after HMDS grafting, the intensity of the ν(OH) absorption bands is strongly reduced. Moreover, new bands appear in the 2900 cm^{-1} region, characteristic of the ν(CH) stretching vibrations in CH_3 groups, and at 1265 cm^{-1}, assigned to the $\delta(CH_3)$ bending vibration in Si-CH_3 groups. These bands are the obvious proof of the surface modification by $Si(CH_3)_3$ trimethylsilyl groups according to the above reaction. It is worth noting that the 1265 cm^{-1} band is quite intense and therefore it can conveniently be used as a marker of the persistency of the $Si(CH_3)_3$ groups on the surface. On the spectra recorded after HMDS addition at room temperature (not shown), absorption bands in the 3400-3200 cm^{-1} region (57, 59) not only indicate the formation of ammonia but also, its subsequent adsorption on Lewis acid sites at the surface of the titania nanoparticles (28, 60). Adsorbed ammonia is totally eliminated after the desorption step at 673 K. The NH_3 formation and adsorption must be kept in mind in the following discussion as ammonia is a reducing agent which may decrease the thickness of the depletion layer of the n-TiO_2 particles. It is indeed observed a loss of transparency of the sample after the grafting experiment, thus indicating an increase of the free carrier density.

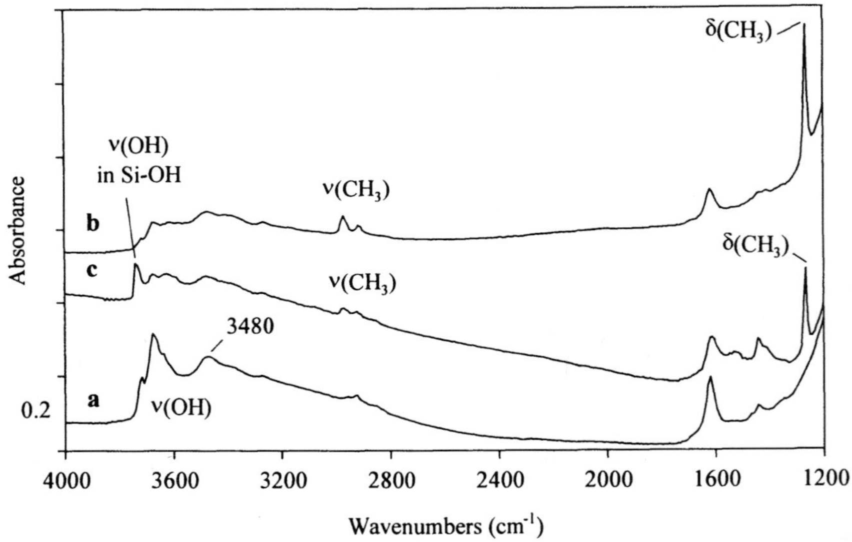

Figure 9. Infrared spectra of the TiO_2 nanopowder recorded at room temperature: a) after activation at 673 K; b) after HMDS grafting at room temperature followed by a desorption at 673 K; c) after exposure to water vapor at 673 K. (The spectra have been shifted for clarity sake).

After modification of whatever surface for a given application, it is important to check whether the chemical groups grafted on this surface will be stable under the operating conditions. In the present case, it must be ensured that the presence of humidity will not ruin the HMDS grafting by a reverse reaction, leading to the re-formation of the OH surface groups. To this end, the HMDS-grafted titania sample was heated at 673 K under water vapor. We observe a very limited recovery of the OH groups (Figure 9c).

But, the most remarkable feature is the appearing of an absorption band at 3738 cm^{-1} which does not exist on the non-grafted sample. This frequency, close to the ν(OH) frequency of the silanol groups on the silica surface (3747 cm^{-1}), is assigned to the ν(OH) stretching frequency of new Si-OH groups (28, 30) formed by hydroxylation of the Ti-O-$Si(CH_3)_3$ grafted groups. Besides, no adsorbed water molecules are observed as the band at 3480 cm^{-1} is not restored. The ν(CH) stretching bands in the 2900 cm^{-1} region and the $\delta(CH_3)$ bending band at 1265 cm^{-1} are still observed, thus proving that the HMDS-grafting resists to a heat-treatment under humidity. It is worth noting that the OH groups which are created by heating under water vapor, exhibit modified properties compared to the original OH groups on the non-grafted n-TiO_2 surface. Indeed, the new formed Si-OH groups should have an acido-basity close to that of the OH groups on a silica surface, that is quite different from the acido-basity of the OH groups on the original titania surface.

5.2. Humidity Effects on Modified Surfaces

The humidity effects on the response of the n-TiO_2 powder toward CO have been presented in section 4.2. In this section, we will evaluate the consequences of the surface modification by HMDS grafting on these humidity effects.

Similar experiments to those described in section 4.2 were performed on the HMDS-grafted n-TiO_2 sample at 673 K. They consisted in the adsorption of four "dry CO" doses (pure CO), followed by the adsorption of four "wet CO" doses (mixture of CO and water vapor), then followed by the adsorption of four new "dry CO" doses. The curve showing the variations of the infrared energy transmitted by the grafted n-TiO_2 powder versus gas exposures is presented in Figure 10. In parallel, the infrared spectra were recorded and the evolutions of the surface species during an experimental step are emphasized by the difference spectra, as explained in section 4.2 (Figure 11).

As already observed, the adsorption of the first "dry CO" doses leads to a reduction which is stronger for the very first dose. The formation of CO_2 is hardly visible on the infrared spectrum (Figure 11a). It is interesting to note that, unlike the CO adsorption on the non-grafted TiO_2 surface (Figures 5 and 7), the overall effect of these first four "dry CO" doses appears as an oxidation. This apparent oxidizing effect could be explained by a re-arrangement of the surface after the reducing effect due to the HMDS grafting (cf. section 5.1). The adsorption of "wet CO" doses amplifies the oxidizing effect which however shows a trend toward stabilization after the second "wet CO" dose. On the infrared spectrum (Figure 11b), we note the formation of a new absorption band at 3730 cm^{-1}.

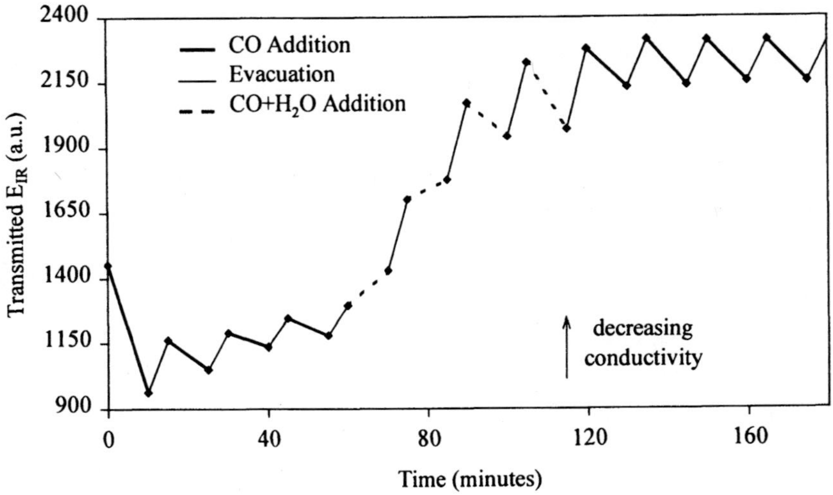

Figure 10. Variations of the infrared energy (E_{IR}) transmitted by the HMDS-grafted TiO_2 nanopowder at 673 K versus gas exposures.

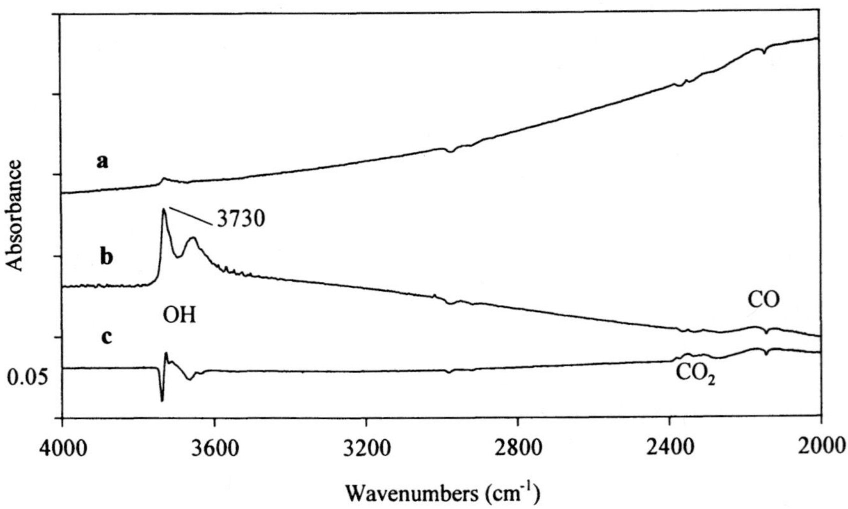

Figure 11. Difference spectra showing the modifications of the HMDS-grafted n-TiO_2 surface at 673 K: a) under CO adsorption (first series of "dry CO" doses); b) under (CO+H_2O) adsorption ("wet CO" doses); c) under CO adsorption (second series of "dry CO" doses). (The difference spectra have been shifted for clarity sake)

This band can be correlated to the one which appeared when the HMDS-grafted sample was heated under water vapor (3738 cm^{-1} at room temperature) (cf. section 5.1) and which was assigned to the ν(OH) stretching vibration in newly formed Si-OH surface groups. Therefore, by analogy, we can conclude that this 3730 cm^{-1} band reveals the formation of Si-OH groups on the n-TiO_2 surface. This formation results in changes in the semiconductor work function and water molecules can act as oxidizing agent. But, after the second "wet CO" dose, the infrared spectra (not shown) indicate that Si-OH groups are no longer formed and then humidity becomes a reducing environment.

When "dry CO" doses are added again, we observe a perfectly reproducible response and no drift of the baseline (Figure 10). On the infrared spectrum (Figure 11c), the formation of CO_2 is more visible and we also observe reproducible and reversible perturbations of the ν(OH) absorption range. The reasons for the apparent stability of the response toward CO in an oxygen-free environment are not yet totally understood. Nevertheless, the importance of the surface chemical composition on the reliability of the sensing properties is clearly established.

6. CONCLUSION

In this chapter, we have presented results proving that Fourier transform infrared spectroscopy is particularly relevant not only for the surface analysis of nanosized particles but also for the study of the electrical properties of semiconducting nanoparticles. A direct correlation of the chemical reactions taking place at the very surface with the resulting variations of the electrical conductivity can be established in real time, which is an important asset for the fundamental understanding of the gas detection mechanism and consequently for optimization of chemical gas sensors based on semiconductors. In addition, the evolution of the newly formed surface chemical species versus time can give information on the stability and reversibility of the gas sensors, thus allowing the determination of the optimum operating conditions.

It has also been demonstrated that specific properties of nanosized particles can be modified on purpose by controlling the surface composition and the surface chemistry. The chemical modification of the surface species on semiconducting nanoparticles generates not only changes in the surface reactivity but also changes in the work function, that is changes in the electrical conductivity and in the sensing properties. Therefore, by tuning the chemical composition of the surface of semiconducting nanoparticles, it can be envisaged to increase the sensor sensitivity and to tailor the sensor selectivity for targeted applications.

7. REFERENCES

1. Somorjai, G.A., *MRS Bulletin* **23**(5), 11 (1998).
2. Somorjai, G.A., *Introduction to Surface Chemistry and Catalysis*, Wiley, New York, 1990.
3. Herzberg, G., *Molecular Spectra and Molecular Structure*, Van Nostrand, Princeton, 1962.
4. Wilson, E.B.Jr, Decius, J.C., and Cross, P.L., *Molecular Vibrations. The Theory of Infrared and Raman Vibrational Spectra*, Dover, NewYork, 1955.
5. Schrader, B. (ed.), *Infrared and Raman Spectroscopy. Methods and Applications*, VCH, Weinheim, 1995.
6. Griffiths, P.R. and de Haseth, J.A., *Fourier Transform Infrared Spectrometry*, Wiley, Chichester, 1986.
7. Baraton, M.-I. in *Handbook of Nanostructured Materials and Nanotechnology*, H.S. Nalwa (ed.), Academic Press, San Diego (1999), pp. 89-153.
8. Baraton, M.-I. in *Nanostructured Materials*, G.M. Chow *et al.* (eds.), NATO-ASI Series, Kluwer Academic Publishers, Dordrecht (1998), pp. 303-317.
9. Baraton, M.-I., *J. High Temp. Chem. Processes* **3**, 545 (1994).
10. Boehm, H.-P. and Knözinger, H. in *Catalysis*, J.R.A. Anderson and M. Boudart (eds.), Springer-Verlag, Berlin (1983), Vol. 4, pp. 39-207.
11. Hair, M.L., *Infrared Spectroscopy in Surface Chemistry*, M. Dekker, New York, 1967.
12. Knözinger, H., *Advances in Catalysis* **25**, 184 (1976).
13. Davydov, A.A., *Infrared Spectroscopy of Adsorbed Species on the Surface of Transition Metal Oxides*, John Wiley & Sons, New York, 1984.
14. Busca, G., Lorenzelli, V., Porcile, G., Baraton, M.-I., Quintard, P., and Marchand, R., *J. Mater. Chem. Phys.* **14**, 123 (1986).
15. Busca, G., Lorenzelli, V., Baraton, M.-I., Quintard, P., and Marchand, R., *J. Mol. Struct.* **143**, 525 (1986).
16. Ramis, G., Busca, G., Lorenzelli, V., Baraton, M.-I., Merle-Mejean, T., and Quintard, P. in *Surfaces and Interfaces of Ceramic Materials*, L.C. Dufour *et al.* (eds.), Kluwer Academic Publishers, Dordrecht (1989), pp. 173-184.
17. Baraton, M.-I., *Nanostruct. Mater.* **5**, 179 (1995).
18. Baraton, M.-I., Chang, W., Kear, B.H., *J. Phys. Chem.* **100**, 16647 (1996).
19. Baraton, M.-I., Merle, T., Quintard, P., and Lorenzelli, V., *Langmuir* **9**, 1486 (1993).
20. Lercher, J., Gründing, C., and Eder-Mirth G., *Catal. Today* **27**, 353 (1996).
21. Lavalley, J.C., *Catal. Today* **27**, 377 (1996).
22. Harrick, N.J., *Internal Reflection Spectroscopy*, Interscience, Wiley, New York, 1967 (2nd printing by Harrick Scientific Corp., Ossining, N.Y., 1979).
23. Gibson, A.F., *J. Scientific Instruments* **35**, 273 (1958).
24. Harrick, N.J., *Phys. Rev.* **125**(4), 1165 (1962).
25. Chabal, Y.J., *Surf. Sc. Reports* **8**, 211 (1988).
26. Baraton M.-I., Merhari L., Chancel F., and Tribout J. in *MRS Symp. Proc.*, MRS Publisher, Warrendale (1997), Vol. 448, pp. 81-86.
27. Morrison, S.R. in *Semiconductors Sensors*, S.M. Sze (ed.), John Wiley & Sons, New York (1994), pp. 383-413.
28. Busca, G., Saussey, H., Saur, O., Lavalley, J.-C., and Lorenzelli, V., *Applied Catal.* **14**, 245 (1985).
29. Primet, M., Pichat, P., and Mathieu, M.-V., *J. Phys. Chem.* **75**(9), 1216 (1971).
30. Morterra, C., *J. Chem. Soc., Faraday Trans. I* **84**(5), 1617 (1988).
31. Ho, S.-W., *J. Chinese Chem. Soc.* **43**, 155 (1996).
32. Tsyganenko, A.A. and Filimonov, V.N., *Spectr. Letters* **5**(12), 477 (1972).
33. Primet, M., Pichat, P., and Mathieu, M.-V., *C. R. Acad. Sc. Paris* **267B**, 799 (1968).

34. Yates, D.J.C., *J. Phys. Chem.* **65**, 746 (1961).
35. Morrow, B.A. in *Spectroscopic Characterization of Heterogeneous Catalysis*, J.L.G. Fierro (ed.), Elsevier, Amsterdam (1990), Part A, pp. A161-A224.
36. Primet, M., Pichat, P., and Mathieu, M.-V., *J. Phys. Chem.* **75**(9), 1221 (1971).
37. Riehemann, W. in *MRS Symp. Proc.,* K.E. Gonsalves, M.-I. Baraton *et al.* (eds.), MRS Publisher, Warrendale (1998), Vol. 501, pp. 3-13.
38. Thornton, E.W. and Harrison, P.G., *J. Chem. Soc. Faraday Trans. I* **71**, 461 (1975).
39. Harrison, P.G. and Maunders, B.M., *J. Chem. Soc. Faraday Trans. I* **80**, 1341 (1984).
40. Cox, D.F., Fryberger, T.B., and Semancik S., *Phys. Rev. B* **38**(3), 2072 (1988).
41. Tribout, J., Chancel, F., Baraton, M.-I., Ferkel, H., and Riehemann, W. in *MRS Symp. Proc.,* K.E. Gonsalves, M.-I. Baraton *et al.* (eds.), MRS Publisher, Warrendale (1998), Vol. 501, pp. 95-100.
42. Barsan, N., Schweizer-Berberich, M., and Göpel, W., *Fresenius J. Anal. Chem.* **365**, 287 (1999).
43. Shimizu, Y. and Egashira, M., *MRS Bull.*, **24**, 18 (1999).
44. Lantto, V. and Romppainen, P., *Surface Science* **192**, 243 (1987).
45. Che, M. and Tench, A.J., *Advances Catal.* **32**, 1 (1983).
46. Henrich, V.E. and Cox, P.A., *The Surface Science of Metal Oxides*, Cambridge University Press, Cambridge, 1994.
47. Clifford, P.K., *Mechanisms of Gas Detection by Metal Oxide Surfaces*, Ph.D. Thesis, Carnegie Mellon Univ., Pittsburg, 1981.
48. Harrison, P.G. and Willett, M.J., *Nature* **332**(6162), 337 (1988).
49. Willett, M.J. in *Techniques and Mechanisms in Gas Sensing,* P.T. Mosley, J.W.O. Norris, and D.E. Williams (eds.), Adams Hilger, Bristol (1991) pp. 61-107.
50. Busca, G. and Lorenzelli, V., *Mater. Chem.* **7**, 89 (1982).
51. Williams, G. and Coles, G.S.V., *J. Mater. Chem.* **8**, 1657 (1998).
52. Many, A., Goldstein, Y., and Grover, N.B., *Semiconductor Surfaces*, North-Holland Publishing Co, Amsterdam, 1965.
53. Morrison, S.R., *The Chemical Physics of Surfaces*, Plenum Press, New York, 1990.
54. Bruening, M., Cohen, R., Guillemoles, J.F., Moav, T., Libman, J., Shanzer, A., and Cahen, D., *J. Amer. Chem. Soc.* **119**, 5720 (1997).
55. Hertl, W. and Hair, M.L., *J. Phys. Chem.* **75**(14), 2181 (1971).
56. Stark, F.O., Johannson, O.K., Vogel, G.E., Chaffee, R.G., and Lacefield, R.M., *J. Phys. Chem.* **72**, 2750 (1968).
57. Chancel, F., Tribout, J., and Baraton, M.-I., *Key Eng. Mater.* **136**, 236 (1997).
58. Baraton, M.-I., Chancel, F., and Merhari, L., *Nanostruct. Mater.* **9**, 319 (1997).
59. Chancel, F., Tribout, J., and Baraton, M.-I. in *MRS Symp. Proc.,* K.E. Gonsalves, M.-I. Baraton *et al.* (eds.), MRS Publisher, Warrendale (1998), Vol. 501, pp. 89-94.
60. Tsyganenko, A.A., Pozdnyakov, D.V., and Filimonov, V.N., *J. Mol. Struct.* **29**, 299 (1975).

NANOCRYSTALLINE OXIDES FOR GAS SENSING

E. Traversa, O. Schäf *, E. Di Bartolomeo, P. Knauth *
Dipartimento di Scienze e Tecnologie Chimiche, Università di Roma "Tor Vergata", 00133 Roma, Italy.
**Université de Provence-CNRS, Laboratoire Matériaux Divisés, Revêtements, Electrocéramiques (MADIREL, UMR 6121), Centre St Charles, Case 26, 13331 Marseille Cedex 3, France*

1. INTRODUCTION

Growing concerns regarding environmental pollution by toxic gases, such as CO, SO_x and NO_x, and global warming by the greenhouse effect, due to CO_2 and CH_4, have triggered the interest in gas sensors [1]. Their main applications can be found for monitoring emissions and provide feedback to reach, for example, an optimum conversion efficiency and reduced emissions in combustion engines for automobiles [2]. So-called "intelligent" microsystems combine an input microsensor, capable of detecting small changes of gas concentrations and transducing them into an electrical signal, a microelectronic data acquisition, storage and treatment unit and an output actuator that can react and modify the current status of the system, for example by adjusting the fuel injection into an automotive combustion engine.

As attractive as solid-state gas sensors appear conceptually, they must satisfy certain requirements according to a given application, including the "3S" sensitivity, selectivity and stability, but also other important criteria such as short response time, easy processing, long lifetime and low cost. In the framework of the development of microtechnologies and equipment compatible with microelectronics, device miniaturization is another important requirement. Selectivity is a particular important issue, because it limits the usefulness of a gas sensor in the presence of varying concentrations of other gases.

Depending on the type of sensor developed (potentiometric, resistive or capacitive), different approaches are possible for the development of innovative and improved gas sensing materials. One approach is to search for "new" materials, especially with the concept of "molecular recognition".

The basic idea is to use specific interactions between the molecule to be detected in the gas phase and a mobile ion in a solid ionic or mixed conductor, for example between NH_3 molecules and Ag^+ ions in AgCl [3, 4] or Cu^+ ions in CuBr [5, 6]. New materials combinations have also shown promising results, such as the use of oxide electrodes combined with ionic conductors in solid-state potentiometric sensors [7, 8]. Using heterojunctions between p- and n-type semiconducting oxides allows to develop new sensing mechanisms for gas and humidity sensors [9].

A particularly important approach is to optimize the materials microstructure in order to improve their gas sensing properties, especially by using fine-grained samples. For semiconductor-type gas sensors, the use of materials with a mean particle size well below 50 nm significantly improved the gas sensing properties [10], based on the following points. First, a large active surface area is assumed to enhance the materials sensitivity. Second, a faster response is expected due to short diffusion paths and rapid grain boundary diffusion. Third, the space charge region thickness can be extended to the whole grain size, leaving grains wholly depleted of charge carriers and improving thus the sensitivity. Fourth, given the lower processing temperature, metastable phases with improved sensing properties, such as better selectivity, can be obtained. Furthermore, a narrow particle size distribution can be achieved that can improve the sintering process of ceramic sensors. These points should be explicitly discussed in each individual case.

However, nanocrystalline materials in gas sensors have also clearly foreseeable drawbacks, especially an unstable microstructure with a tendency to grain coarsening at moderate temperature. This can be a serious issue during long-time use, leading to aging and stability problems and ultimately limiting the lifetime of such sensors.

The objective of this article is to provide a fairly broad, but not exhaustive, overview on recent developments in the domain of nanocrystalline gas sensors. One can subdivide these sensors into three main types: i) equilibrium and mixed potential sensors, ii) bulk conductivity and iii) boundary conductivity sensors [11]. The focus of this article is on semiconductor sensors, where most of the recent work was performed. In this domain, where a great number of materials have been used occasionally, it seemed reasonable to restrict this report to the most studied compounds, including SnO_2, TiO_2, Fe oxides, ZnO, WO_3, Ga_2O_3, and perovskite-type oxides ($LaFeO_3$ and $SmFeO_3$). In order to check if the positive predictions concerning improved sensor properties are matched and how far degradation problems are encountered in the practice, we focus on studies where a comparison of nanocrystalline and microcrystalline materials was made.

2. EQUILIBRIUM AND MIXED POTENTIAL GAS SENSORS

Potentiometric sensors are the oldest type of electrochemical sensors. The classical example of an equilibrium potential sensor is the solid-state oxygen sensor, developed following an initial report by Kiukkola and Wagner [12, 13], produced in millions and installed in most automobiles today. It is based on calcia- or yttria-stabilized ZrO_2 (YSZ), a solid oxygen ion conductor, and platinum electrodes [14, 15]. The cell can be represented by:

Working electrode $O_2(P_{mes})$, Pt | YSZ | Pt, O_2 (P_{ref}) **Reference electrode** (1)

P_{mes} is the oxygen pressure to be determined and P_{ref} is a stable reference pressure, most conveniently air (P_{ref} = 0.21 bar), but metal/metal oxide mixtures can also be used to fix the oxygen partial pressure, which is then calculated from the equilibrium constant of the reaction:

$$Me + \tfrac{1}{2} O_2 \Leftrightarrow MeO \qquad (2)$$

The cathode reaction in cell (1) is:

$$\tfrac{1}{2} O_2 + 2\,e^- \Leftrightarrow O^{2-} \qquad (3)$$

The heterogeneous equilibrium condition can be written using the chemical potential μ for uncharged species and the electrochemical potential η for charged species :

$$\eta(O^{2-}) = 2\,\eta(e^-) + \tfrac{1}{2}\,\mu(O_2) \qquad (4)$$

$$\eta(O^{2-}) = 2\,\mu(e^-) - 2\,F\,\phi + \tfrac{1}{2}\,\mu^\circ(O_2) + \tfrac{1}{2}\,RT\,\ln(P(O_2)/P^\circ) \qquad (5)$$

ϕ is the internal electrical potential and μ° the standard chemical potential. R, T and F are the gas constant, absolute temperature and Faraday's constant, respectively. Similar expressions can be written for the anode side. The electrochemical potential of the oxygen ions in the solid ionic conductor, the chemical potential of the electrons in the metallic conductor and the standard potential of oxygen drop out by subtraction. Therefore, the EMF of this cell follows Nernst's equation:

$$E = (RT/4F)\,\ln P(O_{2,\,mes})/P(O_{2,\,ref}) \qquad (6)$$

This derivation shows that important assumptions are the constancy of the electrochemical potential of the oxygen ions, in other words, a sufficiently high oxygen ion conductivity in the solid electrolyte and that the equilibria are established at the temperature of measurement, in other words that the exchange current density is sufficient.

Many other equilibrium potential sensors can in principle be designed [16, 17]. They can be classified, as firstly proposed by Weppner [18], according to the relation between the target gas and the mobile ion in the solid electrolyte. In so called "type I" sensors, the gas and the mobile ion in the solid electrolyte are related by a direct reaction, such as eq. (3). In "type II" sensors the ions related to the target gas are present, but immobile in the solid electrolyte. Gauthier and Chamberland [19] proposed the first potentiometric sensor of this kind, showing that oxyacid salts could serve as solid electrolyte membranes for oxidic gas sensors. For example, K_2CO_3 was used as solid K^+ ionic conductor for the target gas CO_2. The main problem of this kind of sensor is the lack of thermal and chemical stability of oxyacid salts. In "type III" sensors, by far the most important group, an auxiliary phase (usually oxyacid salts) establishes the relation of the solid electrolyte with the target gas[20]. This allows a larger flexibility for these sensors that are able to detect a wider variety of gases. For example, sensors based on Na^+ ion conductors, such as NASICON [21], are able to detect CO_2, NO_x or SO_2, using the appropriate auxiliary phase [22]. Further recent developments foresee the use of metal oxide electrodes as auxiliary phases in substitution of the oxyacid salts [23, 24, 25]. For these sensors the observed EMF is not described by Nernst's law. Electrochemical gas sensors based on YSZ with metal oxide electrodes, so-called mixed potential sensors, have also been proposed [26, 27, 28, 29]. These non-equilibrium sensors are actually modifications of the well-established oxygen sensor and their technological transfer might be easy.

Although the use of nanocrystalline materials may in principle improve the ionic conductivity of solid electrolytes [30, 31, 32], this possibility is certainly more than counterbalanced by the reduced long-term stability of the nanocrystalline material. Conceptually more interesting is the possibility to enhance the exchange current density, which is determined by the charge transfer kinetics at the solid electrolyte/electrode interface, and thus the reversibility of the cell by using nanocrystalline materials.

For the non-equilibrium sensors, most often several electrode reactions are observed simultaneously. In this case, so-called mixed potential sensors are obtained [33, 34]; mixed potentials depend on the kinetics of the different electrode processes involved. They depend thus strongly on kinetic parameters and, as such, on the electrocatalytic activity of the electrode material. The use of nanocrystalline materials can in principle improve the function of a mixed potential sensor by providing supplementary active surface and grain boundary sites, though very limited investigations have been reported in the literature.

A recent example is about NO_x sensors based on YSZ or NASICON, a solid Na^+ ion conductor, coupled with a rare-earth perovskite $LaFeO_3$ [35]. The nanocrystalline oxide powders were prepared using a chemical route [36]. The grain size of the oxide electrode had an important effect on the gas sensitivity. The response to NO_2 was significantly improved by using nanosized oxide grains. The sensing mechanism might be explained with electrochemical reactions taking place at the triple phase boundary between gas, metal electrode, and solid electrolyte. Though more investigations are necessary to better understand the role of the oxide electrode, $LaFeO_3$ might increase the number of gas adsorption sites and catalyze the electrochemical reactions.

3. SEMICONDUCTOR BULK AND BOUNDARY GAS SENSORS

Semiconductor gas sensors can be subdivided into bulk and surface types. In the first case, the bulk stoichiometry of the material is changed due to interaction with the gas phase. It is clear that this kind of sensor works at higher temperatures. The thermodynamic description of the defect equilibria is state-of-the-art. In the second type, semiconductor surface sensors, gas adsorption phenomena lead to a change of the surface conductivity of the materials, which is used for detection. As no bulk diffusion is necessary, but only surface reactions occur, this type of sensor works normally at lower temperatures than the bulk sensor. The main part of recent literature on the use of nanocrystalline oxides is related to semiconductor surface sensors.

3.1. Bulk conductivity sensors

The variation of the bulk conductivity of semiconducting oxides can be used for gas detection (mainly oxygen). These sensors work at elevated temperatures (typically above 600°C), because diffusion needs a high activation energy. In the case of oxygen, its diffusion into or out of the oxide changes the oxygen bulk stoichiometry, by well-known red-ox reactions. An oxygen uptake compensates an original oxygen deficiency, which is typical for many n-type semiconducting oxides, such as SnO_{2-x}. The kinetics of oxygen stoichiometry changes in n-type semiconducting tin dioxide was recently investigated[37]. The equilibrium can be written using the Kröger-Vink nomenclature :

$$O_O \Leftrightarrow V_O^{\bullet\bullet} + \tfrac{1}{2}\, O_2\,(g) + 2\,e' \qquad (7)$$

The corresponding equilibrium constant can be written:

$$K = [V_O^{\bullet\bullet}][e']^2\, P(O_2)^{1/2} \qquad (8)$$

Square brackets represent concentrations. In absence of significant amounts of doping impurities, the bulk electroneutrality equation can be formulated as:

$$2[V_O^{\bullet\bullet}] = [e'] \qquad (9)$$

After insertion of eq. (9) into eq. (8), one obtains:

$$[e'] = (2K)^{1/3}\, P(O_2)^{-1/6} \qquad (10)$$

The oxygen partial pressure exponent takes the typical value –1/6, which can be found from the oxygen partial pressure dependence of the total bulk conductivity, which is essentially electronic, given the much larger electron mobility μ_e:

$$\sigma = F\mu_e\,(2K)^{1/3}\, P(O_2)^{-1/6} \qquad (11)$$

In principle, the selectivity of bulk conductivity sensors is high, because although several species can be exchanged at the surface of the oxide, only few possess a sufficient bulk diffusivity. Similar bulk conductivity sensors for other gases than oxygen can be conceived. In all cases, a redox reaction such as eq. (7) has to be formulated and a state-of-the-art thermodynamic treatment ends up with a power law similar to eq. (11).

The use of nanocrystalline materials can be of interest for a bulk conductivity sensor, because the large interface density can significantly improve the diffusivity by interface paths. On the other hand, the selectivity might be lower, because diffusion of other species might also be enhanced. Furthermore, the typical power dependence can also be changed in some cases, due to defect interactions, as we will see in the following.

Another typical material for this type of sensor is n-type crystalline TiO_2 (rutile or anatase phase). Knauth and Tuller [38] reported for temperatures between 450 and 580°C that the oxygen partial pressure dependence of the conductivity of nanocrystalline anatase under reducing conditions is much steeper than for the microcrystalline counterpart (Figure 1). This is obviously of interest for bulk sensor applications. A clear difference between coarse-grained and ultra-fine grained samples appears over the whole range of oxygen partial pressures. Here, one must take into account that the coarse-grained samples, obtained after heat treatment at 1400°C, present the thermodynamically stable rutile structure, whereas the nanocrystalline materials are phase-pure anatase. The conductivity dependence can be related to the defect chemistry of the materials.

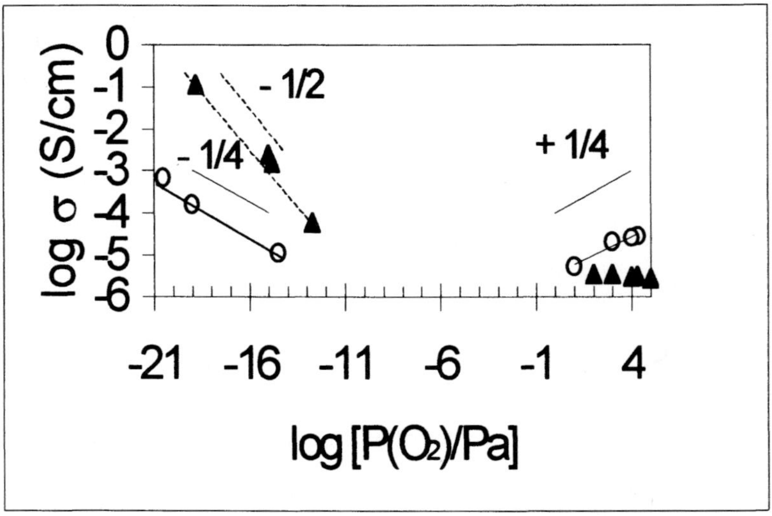

Figure 1. Oxygen partial pressure dependence of the bulk conductivity of nanocrystalline TiO_2 (anatase, D = 35 nm, triangles) in comparison with microcrystalline samples (rutile, D = 1 μm, open dots) at 580°C. From reference [38].

In accordance with older data, the microcrystalline sample shows a regime of p-type conductivity at high $P(O_2)$ with a $P(O_2)^{1/4}$ dependence and a regime of n-type conductivity at low $P(O_2)$ with a $P(O_2)^{-1/4}$ dependence. These power laws can be interpreted by assuming that the reduction of TiO_2 proceeds by the formation of fully ionized titanium interstitials (metal excess: $Ti_{1+x}O_2$) and electrons at low oxygen partial pressure and oxidation by annihilation of interstitials and formation of electron holes at high oxygen partial pressure (metal deficiency: $Ti_{1-x}O_2$). The electroneutrality condition ($4\ [Ti_i^{4\cdot}] = [A_{Ti}'] = const$) assumes control by background acceptor impurities (A, likely sodium in this case).

The nanocrystalline material shows a distinctly different oxygen pressure dependence of conductivity. At high oxygen partial pressures, a conductivity plateau is found and at low $P(O_2)$ a steeper increase of conductivity than that observed for the coarse-grained material is obtained. The power law exponent at low $P(O_2)$ takes the uncommon value of -1/2 in the nanocrystalline material. To understand this result, the defect chemistry in nanocrystalline TiO_2 must be re-evaluated [38, 39].

The reduction-controlled behavior at low oxygen partial pressures can be interpreted by assuming that titanium interstitials are not completely ionized, i.e. that we have association between the highly concentrated defects ($Ti_i^{4\cdot}$ and e'). This condition becomes more likely as the level of nonstoichiometry increases. Under the assumption that doubly charged titanium interstitials are formed, we write the reduction reaction as:

$$V_i + TiO_2 \Leftrightarrow Ti_i^{\cdot\cdot} + O_2 + 2\,e' \tag{12}$$

where V_i is a vacant Ti interstitial site. The equilibrium constant is then:

$$K' = [Ti_i^{\cdot\cdot}].\,[e']^2\,.\,P(O_2) = K_0'\exp(-\Delta_{red}H'/RT) \tag{13}$$

where $\Delta_{red}H'$ is the enthalpy of reduction of the nanocrystalline material. The electroneutrality condition for acceptor doping is:

$$2\,[Ti_i^{\cdot\cdot}] = [A_{Ti}']. \tag{14}$$

Combining these equations, we obtain:

$$\log [e'] = \log (2K'/[A_{Ti}'])^{1/2} - 1/2 \log P(O_2) \tag{15}$$

The oxygen pressure dependence of conductivity is therefore:

$$\log \sigma = A' - 1/2 \log P(O_2) \tag{16}$$

with the experimentally observed exponent $n = -1/2$.

The strong $P(O_2)$ dependence of conductivity under reducing conditions is obviously of interest for gas sensor applications; furthermore, the redox kinetics can be expected to be significantly enhanced in nanocrystalline TiO_2 and further investigations appear worthwhile.

One should mention that similar results were also reported for dense nanocrystalline CeO_2 ceramics[40] and in a series of papers on nanocrystalline thin films of pure CeO_2 [41], Y- [41] and Sc-doped ZrO_2 and $SrCeO_3$[42]. The latter was used in a hydrogen sensor.

3.2. Boundary conductivity sensors

Low-temperature chemisorption of environmental gases changes the surface conductivity of oxides and can be detected in this way. A typical material is n-type semiconducting SnO_2 that is widely used for reducing gas detection also in commercial devices [43, 44, 45]. A widely accepted sensing mechanism is that negatively charged oxygen adsorbates, such as O_2^-, O^- and O^{2-} cover the surface of the oxide particles in air. In the case of n-type semiconducting oxides, electron-depleted space charge regions are formed to counterbalance the negative surface charge:

$$O_2\,(g) + 2\,e^- \Leftrightarrow 2\,O^-(ads) \tag{17}$$

The width of the space charge regions is a function of the oxygen surface coverage and the bulk electron density. The resistance of a n-type semiconducting oxide is therefore high in oxygen-rich environment, given that a space charge barrier develops at every boundary between grains. When the sensor is exposed to a reducing gas, oxygen adsorbates are consumed and electrons previously trapped are injected into the space charge regions, leading to a decrease of the space charge layer width and a reduction of the Schottky barrier height. The corresponding resistance drop is measured [46]. Since semiconductor gas sensors respond more or less to any reducing gas by this mechanism, they usually suffer from cross-sensitivity and a lack of selectivity. These sensors typically work at temperatures below 600°C.

The nanocrystalline character of the materials permits in principle a lowering of the application temperature relative to the microcrystalline materials. Due to the enhanced surface to bulk ratio, a distinctive sensitivity enhancement is also expected using nanocrystalline oxides, allowing the detection of very low gas concentrations.

Tin dioxide SnO_2 is the most prominent oxide used in semiconductor boundary sensor for reducing gases. Some interesting work has been published in recent years concerning nanocrystalline SnO_2, in thin-film [47, 48, 49, 50, 51] and especially in thick-film form by screen printing [52, 53, 54, 55]. Williams and Coles [56] investigated films with nanocrystalline precursor powders made by laser ablation; Martinelli et al. used chemically processed precursors[57]. Obviously, the mean grain size of the precursors can be tuned by selecting the calcination temperature: a high resolution transmission electron micrograph shows the grain-size of sol-gel processed SnO_2 powder calcined at 650°C (Figure 2). Tests performed using films made from this precursor and, for comparison, samples calcined at 850°C show a clearly improved response to various gases of the material with smaller grains (Figure 3). The difference is particularly clear at the lowest operating temperature, whereas the difference is less remarkable at higher temperature. Recently, Morante and coworkers used in-situ scanning tunneling microscopy to study SnO_2 nanoparticles and observed an increase of surface band-gap in oxidizing atmosphere (NO_2) and a decrease under reducing conditions (CO) [58].

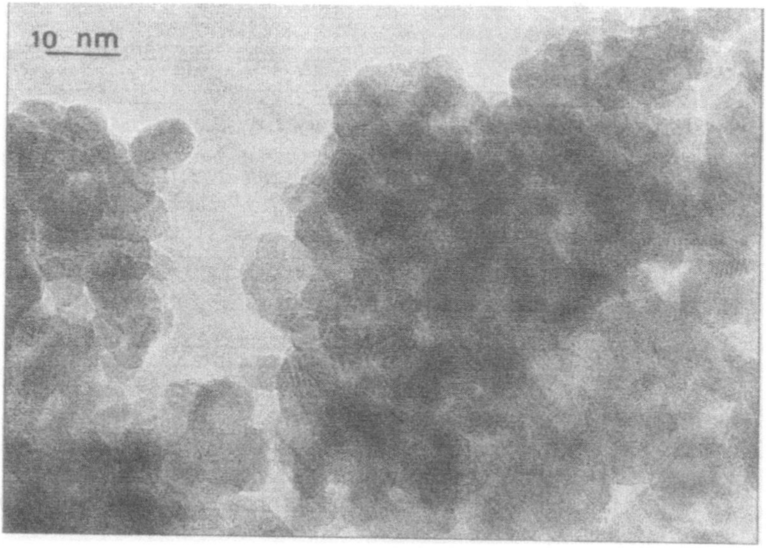

Figure 2. High resolution transmission electron micrograph of a sol-gel processed nanocrystalline SnO_2 powder calcined at 650°C. From reference [57].

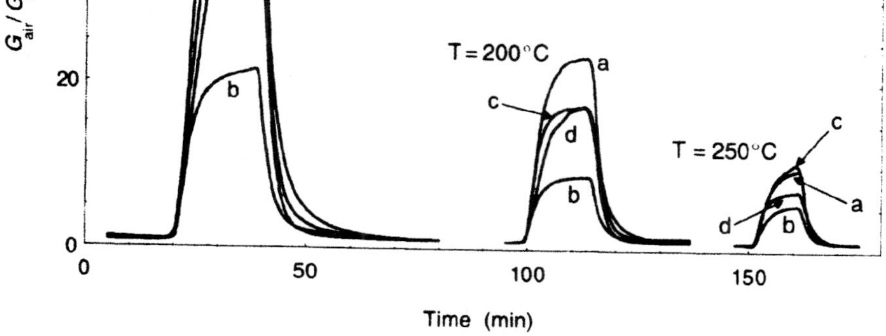

Figure 3. Electrical response at different operating temperatures to NO_2 (10 ppm) in wet air (40% relative humidity). Sol-gel processed pure SnO_2 fired at a) 650°C, b) 850°C. Mo-doped SnO_2 fired at c) 650°C and d) 850°C. From reference [57].

Figure 4 shows another remarkable result. Here, the grain size dependence of the electrical resistance of a SnO_2 sensor in air and in 800 ppm H_2 is plotted (Fig. 4a). One recognizes a drastic change for a grain size below 10 nm[59]. The sensitivity to hydrogen shows also a strong grain size dependence (Fig. 4b).

The qualitative change is probably related to a complete charge carrier depletion of the grains, because the Debye length λ, defined as [3]

$$\lambda^2 = \varepsilon\varepsilon_0 RT/(2F^2[i]) \quad (18)$$

can be estimated to be around 3 nm at the measurement temperature (250°C). [i] is the concentration of electronic charge carriers, while $\varepsilon\varepsilon_0$ is the dielectric permittivity. Space charge regions overlap at $D \leq 4\lambda$ and the grains get fully depleted [3]. It clearly appears from this study that in order to get the full benefit of the "nanocrystallinity", very small particle sizes of the order of the Debye length are necessary. However, these results imply also that supplementary parameters, such as doping impurity concentrations, must be taken into account, because they influence the Debye length by providing mobile charge carriers.

Bunde and co-workers [60] recently analyzed the performance of a nanocrystalline thin layer for gas sensing using a percolation-type model. Here, it is assumed that nanoparticles can be totally depleted of charge carriers and become insulating. The probability to get insulating grains depends on the particle size and the number of neighbor particles. The percolation threshold is attained for a certain critical surface density of reducing gas species. Near this critical concentration, the conductivity change is important, which may be used to improve the sensitivity of nanocrystalline sensors. However, the percolation model assumes a constant grain size and no grain coarsening at operating temperatures.

Other interesting experiments were made on nanocrystalline TiO_2 sensors. Figure 5 shows the microstructure changes observed when the calcination temperature of TiO_2 powders is changed from 650°C to 850°C and when tantalum is added as a dopant, and the corresponding sensor curves for 100 ppm CO in wet air (30% relative humidity) at 400 °C. One recognizes very easily an important influence of the calcination temperature, when no Ta is added. No response is observed without Ta when the sample is heated to 850°C, because grain coarsening is too important. The highest sensitivity is observed with Ta-doped samples [61]. Ta addition prevents grain growth and inhibits the anatase-rutile phase transition; furthermore, the metal is known to be a good catalyst for surface reactions. This behavior was shown not only for TiO_2 powders prepared using sol-gel routes [62, 63], but also for powders prepared using laser pyrolysis [64]. There is also a great amount of work on nanocrystalline TiO_2 in thin-film form. The interested reader is referred to recent literature[65, 66, 67, 68].

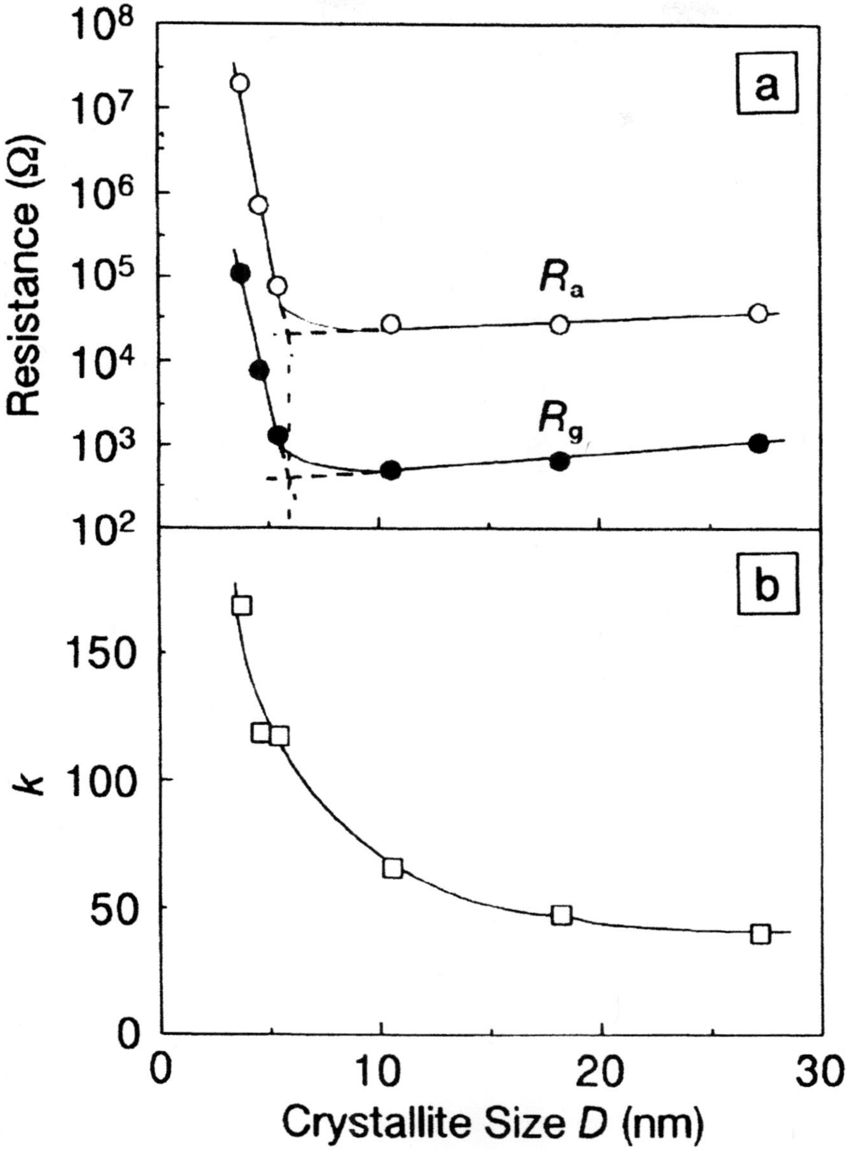

Figure 4. a) SnO_2 sensor resistance in air (R_a) and in 800 ppm H_2 (R_g) and b) sensitivity (k = R_a/R_g) to 800 ppm H_2 as a function of grain size D. From reference [46, 59].

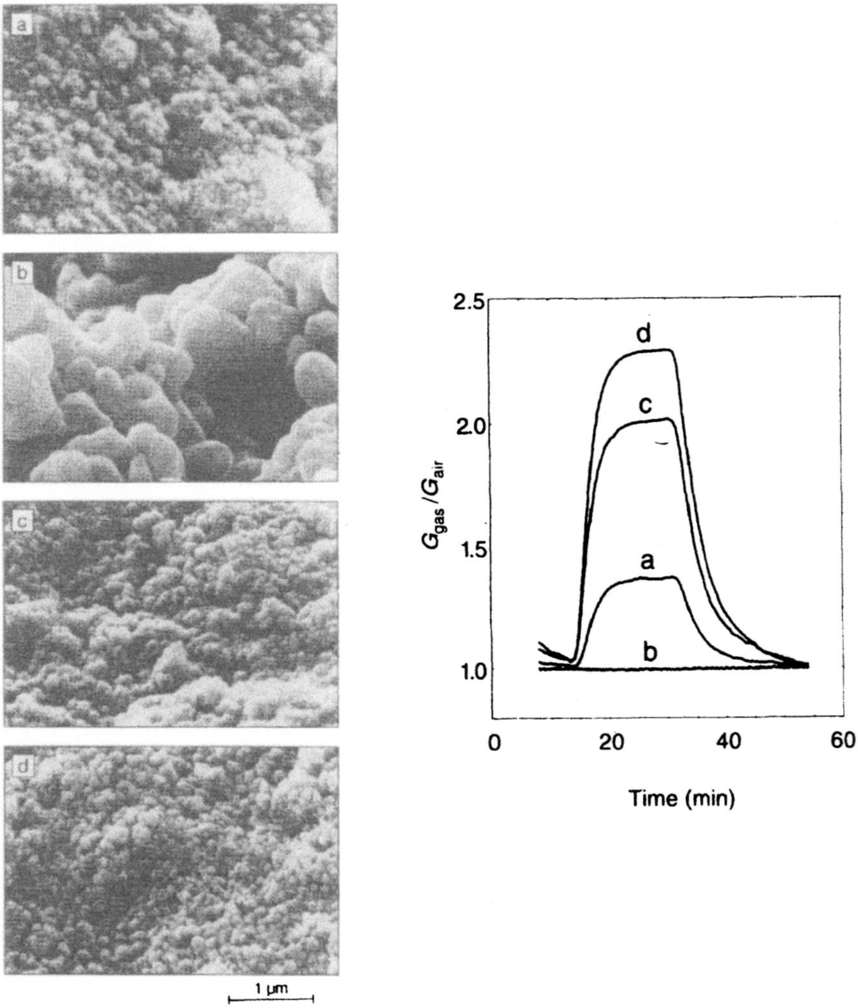

Figure 5. Scanning electron micrographs and electrical responses at 400°C to 100 ppm CO in wet air (30% relative humidity). Pure TiO_2 samples fired at a) 650°C, b) 850°C and 10 at% Ta-doped TiO_2 samples fired at c) 650°C and d) 850°C. From reference [57].

Nanocrystalline rare earth perovskite-type oxides $LaFeO_3$ and $SmFeO_3$ were prepared by thermal decomposition at low temperatures of hexacyano-complexes [69, 70]. These powders were used to produce thick-film gas sensors, fired at various temperatures to modify their grain size [71, 72]. In the case of $SmFeO_3$, some interesting results were found [73, 74]. Surprisingly, the NO_2 response was improved for films with larger grains. This was attributed to different surface states achieved with different processing parameters.

Some of the nanostructured sensors based on SnO_2, TiO_2, and perovskite-type oxides were used for atmospheric pollutant monitoring. The application of nanocrystalline oxides improved the gas sensors' performance in such a way that the difficult determination of extremely low gas concentrations became possible. The interested reader can find details in other recent papers [57, 75, 76, 77].

Nanocrystalline iron oxides for gas sensor applications can be obtained by different processes, chemical vapor deposition of organic precursors [78], decomposition techniques [79, 80], sol-gel techniques [81, 82, 83] or dehydration of γ-FeOOH [84]. As different precursor routes exist, different phases are obtainable: α-Fe(III)oxide (Haematite) has a corundum structure and is the only thermodynamically stable phase, while the γ-phase (Maghemite) is of spinel type and often formed for kinetic reasons.

Liu et al. [78] studied ultrafine Fe_2O_3 powders, tempered at different temperatures (350-800°C) and, therefore, with different grain size and different quantities of the α-, β and γ-phase, using ethanol vapour in synthetic air as test gas. Tianshu et al. [84] investigated the sensing behavior of phase pure γ-iron oxide calcined at 600°C for different combustible gases. Wang et al. [79] used pure α-iron oxide of about 63 nm grain size deposited on Si (111) to detect CH_4 and CO. In all these studies, the maximum sensitivity was found around 400°C. In contrast to these results, Sun et al. [83] found a sensitivity maximum for NO_2 at about 200 °C using α-iron oxide of 50 nm average grain size. Too high sintering temperatures reduced the sensitivity.

Han et al. [80] revealed that an addition of Al to the precursor material preserved the gas response of the sensor material even when the sintering was performed at 850°C, because the doping inhibits grain growth of the nanocrystalline particles. Chauhan et al. [82] investigated the humidity sensitivity of α-iron oxide of 30 nm grain size. The tests were performed below 100°C and the smaller the grain size the larger the humidity response. However, the sensing effect at temperatures above the water boiling point could not be explained satisfyingly.

The conclusion of this literature overview is that although miscellaneous investigations of the sensor properties of different iron oxides have been performed, the studies remain incomplete. Neither a direct comparison between micro- and nanocrystalline materials exists nor the direct influence of crystallinity on sensitivity and selectivity of the sensor signal was investigated. Furthermore, the degree of sintering was not reported and it is not known if the measurements were performed on dense nanomaterials that were already sintered or on loosely compacted samples. The catalytic behavior of the nanomaterials and, in case of γ-Fe_2O_3, the true oxidation state of iron has not yet been investigated. Further systematic studies on the sensing properties and mechanisms of nanocrystalline iron oxides are, therefore, highly desirable.

A comparison between micro- and nanocrystalline ZnO showed a size effect with a strong shift of sensitivity towards lower temperatures [85]. However, an influence of partial disordering caused by the arc plasma preparation method on this behavior could not be excluded. Chadwick and coworkers [86] already noticed that some disorder in this material is quite important for its sensor application. Paraguay et al. studied the influence of dopants on the sensitivity of nanocrystalline ZnO thin-films for ethanol vapour [87].

The high chemical and thermal stability make WO_3 an interesting material for long time applications under rough conditions. Capone et al. [88] showed that nanocrystalline films of WO_3 exhibit a better NO sensing characteristics than films of vanadium oxide V_2O_5 [89] with a sensitivity maximum at 200°C and a much faster response. Using ethanol and other combustible gases, W.Yu-De et al. [90] found a strong influence of grain size on the sensitivity. These results are comparable to those of Solis et al. [91] who found at room temperature, a maximum of sensitivity at calcination temperatures around 500°C. Together with a grain size effect, a phase transition is suggested to be responsible for this behavior with both phases (monoclinic and tetragonal) present at the sensitivity maximum.

Composite materials used in conductometric sensors are mostly dispersions of second phase particles in the bulk or on the surface of the main gas sensitive material. If the second phase is a metal, one may define the mixture as composite if the percolation threshold of the metal is not attained. Second phase particles can have catalytic, electroactive and size-stabilizing effects [92]. Catalytic additives, such as Pt, enhance specific reaction rates at the surface of the gas sensitive material. The catalytic effect of dispersed noble metals has been used to improve the selectivity of gas sensors. Electroactive metals, such as Pd or Ag, with work functions higher than the work function of the n-type oxide can extract electrons from the oxide and, as a result, lead to an enhancement of the Debye length and the sensitivity of the sensor. Stabilizing additives, such as Nb, Al or Si, inhibit grain growth. If the second phase is an oxide, it helps to catalyze boundary reactions or stabilizes the morphology of the nanocrystalline material by avoiding grain growth.

The group of Meixner worked initially on microcrystalline oxygen sensors based on Ga_2O_3 [93], but soon extended the investigations to nanocrystalline thin films [94]. Undoped Ga_2O_3 thick- and thin films showed comparable gas sensing properties with respect to sensitivity, reproducibility and stability for O_2, CO and CH_4 [95, 96]. Although different oxide composites were investigated [97], the introduction of dispersed Au particles of nanometer-size gave the best sensitivity and selectivity, especially with respect to CO, in a narrow temperature window at about 550°C [98].

4. CONCLUSIONS

The common objective of a conclusion is to recollect the knowledge gained so far and to indicate new paths for research. Here, the goal will be more modest, taking into consideration that the subject is too recent to get already a sufficiently broad understanding. One of the – rather frustrating – conclusions is that so far a large amount of the sensor work in literature is very phenomenological in nature and that definitely more work is necessary on the fundamentals. However, one can draw a few conclusions on the type of work that should be pursued further and that was highlighted in this article.

- The literature overview confirms that preparation conditions ultimately influence the chemical and crystallographic properties. The investigations in the sensor field have to implement both morphology and electrical properties by engineering bulk and boundary defects.
- Experimental observations of the microstructure, such as those shown in fig. 2 and 5, and its evolution with time are a necessity. They permit to observe the grain size distribution, to check the presence of amorphous residues and other secondary phases, especially in the intergranular regions.
- The comparison of nanocrystalline and microcrystalline materials, obtained by sintering at different temperatures, like in fig. 3, is a necessity, if one wants to verify the assumption that results are related to the reduced grain size of the material and not for example to spurious impurity effects. As straightforward as this operation seems to be, it is too seldom realized in sensor work on nanocrystalline materials.
- The use of nanocrystalline oxides for gas sensors gives the possibility to improve their performance, especially in terms of low levels of gas concentrations that can be detected. This opens a variety of new technological applications for solid-state gas sensors, such as the outdoor monitoring of atmospheric pollutants.
- A goal for future work is to determine more often the fundamental laws with quantitative parameters, such as partial pressure exponents or activation energies, that can be checked with state-of-the-art theory. One such approach on the bulk conductivity of nanocrystalline TiO_2 was highlighted here and is still in progress. The partial pressure dependence of the surface conductivity can also in certain cases be described by simple power laws. But in any case, the discovery of such relations is a precious tool for the planning of further experiments and to predict opportunities for gas sensing.
- The modeling of nanocrystalline gas sensors is also still in its infancy. One problem for application of the models is the mediocre long-term

stability of nanocrystalline materials. The application of ab-initio techniques and computer simulations for calculation of surface properties [99] and gas phase interactions should be continued.

- In the domain of potentiometric sensors, nanocrystalline oxides have been seldom explored to date. Mixed potential sensors, whose potential depend on the kinetics of different concurrent reactions, are certainly a field where the use of nanocrystalline materials in the electrodes could make a difference, because the redox kinetics are sensitive to surface engineering. Especially composites with nanocrystalline metals are worthwhile investigating.
- Surface conductivity sensors use, generally speaking, space charge effects for gas detection. Given the low temperatures of operation, mostly electronic effects are observed, but ionic effects are possible for "acid-base" type sensors [100]. This kind of sensor is thus sensitive to size effects with a linear variation expected as long as the width of space charge regions is significantly smaller than the mean grain size. However, the study highlighted in fig. 4 shows one case where the sensor sensitivity depends non-linearly on the grain size below a critical value, indicating mesoscopic effects, due to overlapping space charge regions. The related strong sensitivity increase shows that the full potential of nanocrystalline oxides for gas sensing is only obtained at very low grain size or, vice versa, for a large Debye length. One way to obtain the latter is to reduce the concentration of all doping impurities. Another is to provide sufficient interface sites for impurity segregation. The recent discovery of mesoscopic heterolayers with high ionic conductivity, prepared by molecular beam epitaxy [101], leads the way.

Gas sensing is one of the domains where the use of nanocrystalline oxides appeared immediately very promising. After the first few years of intense work, some successful and interesting studies have been performed, as reported in the present review. Nevertheless, the real proof of the superiority of nanocrystalline oxides over their conventional counterparts is still awaiting and especially their long-term stability is an open question. Further innovative developments might be foreseen with the use of mesoscopic systems, but the obtainment and use of high-purity, high-quality materials in this domain is yet to come.

ACKNOWLEDGMENT

The authors gratefully acknowledge the support of the European Union in the framework of the COST 525 project: "Grain Boundary Engineering of Nanocrystalline Materials for Solid-State Electrochemical Devices – NanoMASSED".

REFERENCES

[1] H. Baltes, W. Göpel, J. Hesse (Eds). Sensors : A Comprehensive Survey, Vol. 2, VCH, New York, 1996.

[2] E. M. Logothetis, J. Solid State Chem., 12, 331 (1975).

[3] J. Maier, Prog. Solid State Chem., 23, 171 (1995).

[4] J. Maier, U. Lauer, W. Göpel, Solid State Ionics, 40/41, 463 (1990).

[5] M. Bendahan, C. Jacolin, P. Lauque, J.-L. Seguin, P. Knauth, J. Phys. Chem. B, 105, 8327 (2001).

[6] O. Schäf, P. Lauque, J.-L. Seguin, M. Eyraud, P. Knauth, Thin Solid Films, 389, 5 (2001).

[7] E. Di Bartolomeo, E. Traversa, M. Baroncini, V. Kotzeva, R. V. Kumar, J. Eur. Ceram. Soc., 20, 2691 (2000).

[8] T. Hibino, A. Hashimoto, S. Kakimoto, M. Sano, J. Electrochem. Soc., 148, H1 (2001).

[9] E. Traversa, J. Am. Ceram. Soc., 78, 2625 (1995).

[10] C. Xu, J. Tamaki, N. Miura, N. Yamazoe, Sensors and Actuators B, 3, 147 (1991).

[11] J. Maier, in Oxygen Ion and Mixed Conductors and Their Technological Applications, Kluwer, Netherlands (2000), p. 399.

[12] K. Kiukkola, C. Wagner, J. Electrochem. Soc., 104, 379 (1957).

[13] J. Weissbart, R. Ruka, Rev. Sci. Instr., 32, 593 (1961).

[14] A. D. Brailsford, M. Yussouff, E. M. Logothetis, Sensors and Actuators B, 44, 321 (1997).

[15] C. Déportes, M. Duclot, P. Fabry, J. Fouletier, A. Hammou, M. Kleitz, E. Siebert, J.-L. Souquet, Electrochimie des Solides, Presses Univ. Grenoble (1994), p. 343.

[16] N. Yamazoe, N. Miura, MRS Bulletin, 24(6), 37 (1999).

[17] N. Yamazoe, N. Miura, J. Electroceram., 2, 243 (1998).

[18] W. Weppner, Proceedings 2nd International Meeting on Chemical Sensors, Bordeaux, France (1986), p. 59.

[19] M. Gauthier, A. Chamberland, J. Electrochem. Soc., 124, 1579 (1977).

[20] Y. Saito, T. Maruyama, Y. Matsumoto, K. Kobayashi, Y. Yano, Solid State Ionics, 14, 273 (1984).

[21] Y. Saito, T. Maruyama, Solid State Ionics, 28-30, 1644 (1988).

[22] N. Miura, N. Yamazoe, in Sensors Update, Vol. 6, H. Baltes, W. Göpel, J. Hesse (Eds), Wiley-VCH, Weinheim, Germany (2000), p. 191.

[23] S. Bredikhin, J. Liu, W. Weppner, Appl. Phys. A, 57, 37 (1993).

[24] Y. Shimizu, K. Maeda, Sensors and Actuators B, 52, 84 (1998).

[25] Y. Shimizu, N. Yamashita, Sensors and Actuators B, 64, 102 (2000).

[26] N. Miura, H. Kurosawa, M. Hasei, G. Lu, N. Yamazoe, Solid State Ionics, 86-88, 1069 (1996).

[27] K.Y. Ho, M. Miyayama, H. Yanagida, J. Ceram. Soc. Jpn., 104, 995 (1996).

[28] E.L. Brosha, R. Mukundan, D.R. Brown, F.H. Garzon, J.H. Visser, M. Zanini, Z. Zhou, E.M. Logothetis, Sensors and Actuators B, 69, 171 (2000).

[29] J.W. Yoon, M. L. Grilli, E. Di Bartolomeo, R. Polini, E. Traversa, Sensors and Actuators B, 76, 483 (2001).

[30] H. L. Tuller, Solid State Ionics, 131, 143 (2000).

[31] J. Schoonman, Solid State Ionics, 135, 5 (2000).
[32] W. Puin, S. Rodewald, R. Ramlau, P. Heitjans, J. Maier, Solid State Ionics, 131, 159 (2000).
[33] F. Shimizu, N. Yamazoe, T. Seiyama, Chem. Lett., 299 (1978).
[34] N. Miura, G. Lu, N. Yamazoe, Solid State Ionics, 136-137, 533 (2000).
[35] M. L. Grilli, E. Di Bartolomeo, E. Traversa, J. Electrochem. Soc., 148, H98 (2001).
[36] E. Traversa, M. Sakamoto, Y. Sadaoka, J. Am. Ceram. Soc., 79, 1401 (1996).
[37] B. Kamp, R. Merkle, J. Maier, Sensors and Actuators B, 77, 534 (2001).
[38] P. Knauth, H. L. Tuller, J. Appl. Phys., 85, 897 (1999).
[39] P. Knauth, H. L. Tuller, Solid State Ionics, 136-137, 1215 (2000).
[40] Y.-M. Chiang, E. Lavik, I. Kosacki, H. L. Tuller, J. Y. Ying, J. Electroceramics, 1, 7 (1997).
[41] I. Kosacki, T. Suzuki, V. Petrovsky, H. U. Anderson, Solid State Ionics, 136-137, 1225 (2000).
[42] I. Kosacki, H. U. Anderson, Appl. Phys. Lett., 69, 4171 (1996).
[43] N. Taguchi, Japanese Patent Application No. 45-38200 (1962).
[44] K. Ihokura, J. Watson, The Stannic Oxide Gas Sensor- Principles and Applications, CRC Press, Boca Raton, FL (1994).
[45] W. Göpel, K. D. Schierbaum, Sensors and Actuators B, 26, 1 (1995).
[46] Y. Shimizu, M. Egashira, MRS Bulletin, 24(6), 18 (1999).
[47] D.G. Rickerby, M.C. Horrillo, J.P. Santos, P. Serrini, Nanostructured Mater., 9, 43 (1997).
[48] M.J. Willett, V.N. Burganos, C.D. Tsakiroglou, A.C. Payatakes, Sensors and Actuators B, 53, 76 (1998).
[49] A. Rosental, A. Tarre, A. Gerst, T. Uustare, V. Sammelselg Sensors and Actuators B, 77, 297 (2001).
[50] A. Karthigeyan, R. P. Gupta, K. Scharnagl, M. Burgmair, M. Zimmer, S. K. Sharma, I. Eisele, Sensors and Actuators B, 78, 69 (2001).
[51] D-S. Lee, G.-H. Rue, J.-S. Huh, S.-D. Choi, D.-D. Lee, Sensors and Actuators B, 77, 90 (2001).
[52] S.G. Ansari, P. Boroojerdian, S.R. Sainkar, R.N. Karekar, R.C. Aiyer, S.K. Kulkarni, Thin Solid Films, 295, 271 (1997).
[53] G. Williams, G.S.V. Coles, J. Mater. Chem., 8, 1657 (1998).
[54] A. Diéguez, A. Romano-Rodríguez, J.L. Alay, J.R. Morante, N. Barsan, J. Kappler, U. Weimar, W. Göpel, Sensors and Actuators B, 65, 166 (2000).
[55] A. Chiorino, G. Ghiotti, F. Prinetto, M. C. Carotta, C. Malagù, G. Martinelli, Sensors and Actuators B, 78, 89 (2001)
[56] G. Williams, G. S. V. Coles, MRS Bulletin, 24(6), 25 (1999).
[57] G. Martinelli, M. C. Carotta, E. Traversa, G. Ghiotti, MRS Bulletin, 24(6), 30 (1999).
[58] J. Arbiol, P. Gorostiza, A. Cirera, A. Cornet, J. R. Morante, Sensors and Actuators B, 78, 57 (2001).
[59] C. Xu, J. Tamaki, N. Miura, N. Yamazoe, Denki Kagaku, 58, 1143 (1990).
[60] M. Ulrich, C.-D. Kohl, A. Bunde, Thin Solid Films, 391, 299 (2001).
[61] E. Traversa, M. L. Di Vona, S. Licoccia, J. Sol-Gel Sci. Technol., 19, 193 (2000).
[62] M.C. Carotta, M.A. Butturi, G. Martinelli, M.L. Di Vona, S. Licoccia, E. Traversa, Electron Technol., 33, 113 (2000).
[63] E. Traversa, M.L. Di Vona, S. Licoccia, M. Sacerdoti, M.C. Carotta, L. Crema, G. Martinelli, J. Sol-Gel Sci. Technol., 22, 167 (2001).
[64] M. Musci, M. Notaro, F. Curcio, C. Casale, G. De Michele, J. Mater. Res., 7, 2846 (1992).
[65] E. Comini, V. Guidi, C. Frigeri, I. Rocco, B. Sberveglieri, Sensors and Actuators B, 77, 16 (2001).
[66] M. Carotta, M. Ferroni, V. Guidi, G. Martinelli, Adv. Mater. 11, 943 (1999).
[67] G. Sberveglieri, M. Ferroni, V. Guidi, G. Martinelli, P. Nelli, G. Faglia, Nanostruct. Mater., 7, 709 (1996).

[68] M. Ferroni, M. C. Carotta, V. Guidi, G. Martinelli, F. Ronconi, M. Sacerdoti and E. Traversa, Sensors and Actuators B, 77, 163 (2001).
[69] Y. Sadaoka, K. Watanabe, Y. Sakai, M. Sakamoto, J. Alloys and Compounds, 224, 194 (1995).
[70] E. Traversa, P. Nunziante, M. Sakamoto, Y. Sadaoka, M.C. Carotta, G. Martinelli, J. Mater. Res., 13, 1335 (1998).
[71] M.C. Carotta, G. Martinelli, Y. Sadaoka, P. Nunziante, E. Traversa, Sensors and Actuators B, 48, 270 (1998).
[72] G. Martinelli, M.C. Carotta, M. Ferroni, Y. Sadaoka, E. Traversa, Sensors and Actuators B, 55, 99 (1999).
[73] E. Traversa, S. Villanti, G. Gusmano, H. Aono, Y. Sadaoka, J. Am. Ceram. Soc., 82, 2442 (1999).
[74] H. Aono, M. Sato, E. Traversa, M. Sakamoto, Y. Sadaoka, J. Am. Ceram. Soc., 84, 341 (2001).
[75] E. Traversa, Y. Sadaoka, M.C. Carotta, G. Martinelli, Sensors and Actuators B, 65, 181 (2000).
[76] M.C. Carotta, G. Martinelli, L. Crema, M. Gallana, M. Merli, G. Ghiotti, E. Traversa, Sensors and Actuators B, 68, 1 (2000).
[77] M.C. Carotta, G. Martinelli, L. Crema, C. Malagù, M. Merli, G. Ghiotti, E. Traversa, Sensors and Actuators B, 76, 336 (2001).
[78] Y. Liu, W. Zhu, O.K. Tan, Y. Shen, Mat. Sci. Eng. B47, 171 (1997).
[79] S. Wang, W. Wang, Wh. Wang Z. Jiau Y. Qian, Sensors and Actuators B 69, 22 (2000).
[80] J.S. Han, T. Bredow, D.E. Davey, A.B. Yu, D.E. Mulcahy, Sensors and Actuators B 75, 18 (2001).
[81] X.Q. Liu, S.W. Tao, Y.S. Shen, Sensors and Actuators B 40, 161 (1997),
[82] P. Chauhan, S. Annapoorni, S.K. Trikha, Thin Sold Films 346, 266 (1999).
[83] H.-T. Sun, C. Cantalini, M. Faccio, M. Pelino, Thin Solid Films 269, 97 (1995).
[84] Z. Tianshu, L. Hongmei, Z. Huanxing, Z. Ruifang, S. Yushen, Sensors and Actuators B 32, 181 (1996).
[85] L.F. Dong, Z.L. Cui and Z.K. Zhang, Nanostruct. Mater., 8, 815 (1997).
[86] N.V. Russell, A.V. Chadwick, A. Wilson, Nucl. Instr. Methods in Physics Research B, 97, 575 (1995).
[87] F. Paraguay D., M. Miki-Yoshida, J. Morales, J. Solis, W. Estrada L., Thin Solid Films, 373, 137 (2000).
[88] S. Capone, R. Rella, P. Siciliano, L. Vasanelli, Thin Solid Films 350, 264 (1999).
[89] R. Rella, P. Siciliano, A. Cricenti, R. Generosi, M. Girasole, L. Vanzetti, M. Anderle and C. Coluzza, Thin Solid Films, 349, 254 (1999).
[90] W. Yu-De, C. Zhan-Xian, L. Yan-Feng, Z. Zhen-Lai, W. Xing-Hui, Solid State Electronics, 45, 639 (2001).
[91] J.L. Solis, S. Saukko, L. Kish, C.G. Granqvist, V. Latto, Thin Solid Films, 391, 255 (2001).
[92] N. Yamazoe, Sensors and Actuators B, 5, 7 (1991).
[93] M. Fleischer, H. Meixner, Sensor and Actuators B, 4, 437 (1991).
[94] M. Fleischer, H. Meixner, Sensor and Actuators B, 6, 257 (1992).
[95] J. Frank, M. Fleischer, H. Meixner, Sensor and Actuators B, 48, 318 (1998).
[96] T. Schwebel, M. Fleischer, H. Meixner, Sensor and Actuators B, 65, 176 (2000).
[97] J. Frank, M. Fleischer, H. Meixner, A. Feltz, Sensors and Actuators B, 49, 110 (1998).
[98] T. Schwebel, M. Fleischer, H. Meixner, C.-D. Kohl, Sensor and Actuators B, 49, 46 (1998).
[99] G. Balducci, J. Kaspar, P. Fornasiero, M. Graziani, M. S. Islam, J. Phys. Chem. B, 102, 557 (1998).
[100] J. Maier, Sensors and Actuators B, 65, 199 (2000).
[101] H. Sata, K. Eberman, K. Eberl, J. Maier, Nature, 408, 946 (2000).

NANOSTRUCTURED MATERIALS IN PHOTOELECTROCHEMICAL APPLICATIONS

Arie Zaban
Department of Chemistry, Bar-Ilan University, Ramat-Gan 52900, Israel

1. ABSTRACT

Nanostructured materials are used in many photoelectrochemical applications primarily due to surface area considerations. These materials are usually applied as nanoporous electrodes, which are made by the sintering of nanosize semiconductors on a conductive substrate. A surface area factor greater than one thousand produces a significant enhancement of surface processes. However, the porous geometry induces special characteristics that significantly differentiate nanoporous electrodes from their compact analogues. These characteristics must be considered when using these electrodes either in applications or as a research tool. This chapter discusses basic concepts related to the photoelectrochemistry of nanoporous electrodes. A major part of the chapter is devoted to dye sensitized solar cells, one of the most successful examples of the utilization of nanostructured materials in photoelectrochemistry.

2. INTRODUCTION

The use of nanosize semiconductors in photoelectrochemical systems has increased in recent years in parallel to the dramatic development of many branches of nano-technology. The literature provides numerous photoelectrochemistry-related applications in which the special size-induced properties of nanoparticles are utilized. For example, one can find new types or improved versions of photoelectrochemical light emitting diodes[1], photoconductors[2,3], electrochemically assisted photocatalysis, power and photo-electrochromic windows[4-6] and solar cells[7,8].

The application of nanosize materials in photoelectrochemical systems is generally motivated by size dependent properties. These properties can be divided into three main categories: photophysical, structural and physical. The first category relates to the special photophysical properties that are induced, and to some extent, tunable by the size of the material. For example, both the band gap and the bands' position relative to vacuum, become size-dependent below a critical value[9-11]. Such changes usually result in an improved matching of various components in a given system; but, these changes also result in the fabrication of totally new systems. The second category relates to the material structure in terms of purity, stability and activity. For example, studies have found that nanoparticles tend to be more pure than bulk materials[12], and this purity affects nanoparticle performance in operating systems. Phase stability is also highly affected by the size, as the surface tension becomes a major factor in the particle stability. For example, anatase TiO_2 becomes more stable than rutile under 14 nm due to the lower tensions of the anatase structure surfaces[13-16]. These structural properties widen the operation window to more extreme conditions[13,17]. However, apparently the most important reason for the use of nanosize materials in photoelectrochemical applications relates to the third category, and in particular, to the very high surface area that can be achieved when nanoparticles are used to fabricate electrodes[4,18-21]. We refer to what is known as the nanoporous or mesoporous electrodes. These electrodes consist of nanosize semiconductor colloids that are sintered on a transparent conducting substrate. The sintering process forms electrical contact between the various colloids, and between the colloids and the substrate[4,20]. The electrodes have a porous geometry and a very large surface area. For example, when 10-20 nm colloids are used, the surface area of a 10 μm thick electrode is approximately one thousand times greater than the substrate area[22].

The best example of a successful utilization of nanoparticles in a photoelectrochemical application relates to dye sensitized solar cells (also known as Grätzel cells)[7,8,23]. These cells are based on charge transfer processes that occur upon light absorption by dyes that are attached to a semiconductor surface. When compact electrodes were used in dye sensitized cells, very low efficiencies were achieved because of the low absorbance of the sensitizing dye monolayer[24-26]. Increasing the absorbance by attaching additional dye layers did not improve the performance because of the low charge transfer efficiencies of dye multilayers[27,28]. However, with the introduction of nanoporous TiO_2 electrodes in dye sensitized solar cells, the light-to-energy conversion efficiencies of these systems increased from a fraction of a percentage to more than 11%[29,30]. This dramatic improvement in solar-to-energy conversion efficiency is mainly attributed to the large electrode surface area. However, other nanosize-related issues, such as surface states and the absence of space charge, also contribute to the efficiency increase. In any case, the dye sensitized solar cells clearly demonstrate how the use of nanostructured materials can transform an ineffective photoelectrochemical system into an efficient solar cell.

As an attractive system for both basic and applied science, dye sensitized solar cells were investigated intensively in the last decade. Despite being related to dye sensitized solar cells the information obtained by these studies provides the basic understanding for the general field of nanostructured-based photoelectrochemistry. The remainder of this chapter will discuss the dye sensitized solar cell system, emphasizing issues that are related to the role of the nanoporous electrode in this photoelectrochemical cell. We will also describe the operational and constructional concepts of the dye sensitized solar cells.

3. DYE SENSITIZED SOLAR CELL OPERATION

Figure 1 presents a basic energy diagram of the dye sensitized solar cell operation. The basic system contains a wide band gap semiconductor electrode, dye that is attached to the semiconductor, redox electrolyte and a counter electrode (the specific materials are discussed later). Upon illumination of the dye sensitized solar cell, an electron is injected from the dye into the semiconductor film (Figure 1a). Following the injection, a hole is transferred to the redox electrolyte, thus regenerating the dye (Figure 1b). The injected electrons must cross the semiconductor layer and reach the conducting substrate (Figure 1c), while the oxidized ions diffuse towards the counter electrode (Figure 1d) where they are reduced to their original state by the electron traveling through the external wire (Figure 1e)[7]. Consequently, while there is no net change in the system, electrons flow through the external wire.

The processes involved in the operation of the dye sensitized solar cell are quite efficient. In particular, the initial charge separation i.e. the electron injection, is an ultra fast efficient process[31-35]. However, a sufficient optical density that is required for efficient solar energy conversion necessitates the use of high surface area semiconductor electrodes[7,36]. The nanoporous electrodes mentioned above provide the high surface area, thus the operation of the dye sensitized solar cell is better described in the schematic presentation of Figure 2. In this figure the transport issues regarding both the photoinjected electron (Figure 1c) and the electrolyte ions (Figure 1d) are highlighted. Because the major part of this chapter discusses the nanoporous electrode these transport issues will be detailed later. At this point, we note that the transport in the nanoporous region increases the probability of loss mechanisms in the solar cell.

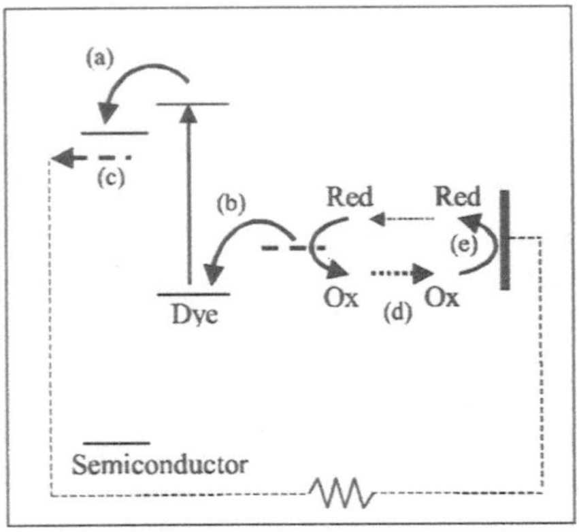

Figure 1. A basic energy diagram of the dye sensitized solar cell operation.

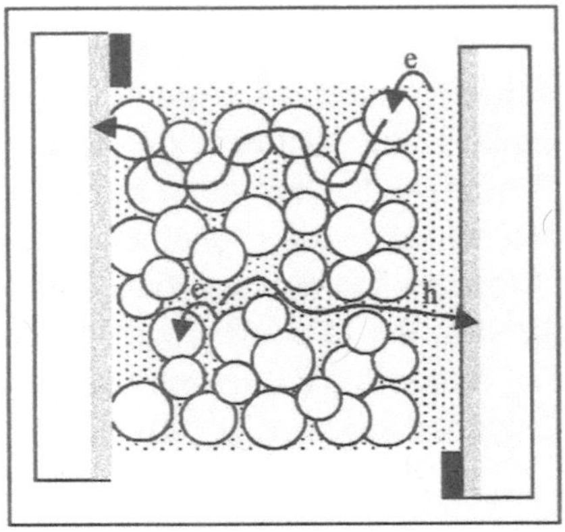

Figure 2. The transport of the photo-injected electrons and the electrolyte ions in the nanoporous film.

3.1. The Injection Process (Figure 1a)

The electron injection from the excited dye to the semiconductor is an ultra fast process occurring in the Pico-second time scale[31-35]. The variance in the reported injection rate values may be attributed to the difference between the measured systems, since the injection rate depends on the specific dye – semiconductor system[37]. The measurements of these ultra fast reactions can be altered by phenomenon such as the surrounding atmosphere, thus providing different values for what can be considered similar systems.

The ultra fast injection rate is explained by a good coupling between the dye and the semiconductor[31-35]. In the efficient dye – semiconductor systems, the injection rate is much faster than the characteristic decay time of the dye, which results in a quantum yield that is close to one. From the point of view of dye design, the ultra fast injection enables the use of dyes whose decay time is relatively short[38]. Furthermore, research has shown that even hot electrons can be transferred in some dye–semiconductor couples[39].

In some systems it was even suggested that the electrons are excited directly into the semiconductor instead of internal excitation followed by injection[40]. It is important to note however, that at this point a comprehensive understanding of the dye–semiconductor system that will enable dye design, has not yet been developed.

3.2. The Regeneration Process (Figure 1b)

The dye regeneration process refers to the electron transfer from the redox electrolyte to the ground state of the oxidized dye. The oxidized ion carries the photo-generated hole to the counter electrode. The dye regeneration is a slow process compared with the injection[41]. The regeneration rate is influenced by both the potential difference between the solution and the ground state of the dye and by the quality of the interaction between the dye and the ions[42,43]. The regeneration rate is a significant factor in dye sensitized solar cells because it determines the average time that a dye is oxidized. An oxidized dye acts as a recombination center that can recapture injected electrons. In addition, most dyes are not stable in their oxidized form. Thus, slow dye regeneration decreases both the efficiency and the stability of a dye sensitized solar cell. Measurements of cells containing the N3 dye and the I^-/I_3^- redox couple (see details in the materials section) found that the regeneration rate in this composition allows approximately 10^8 redox turnovers which corresponds to 15 years of outdoor operation[36].

3.3. The Charge Transport (Figure 1c-d)

The charge transport of both the photo-generated electrons in the semiconductor and the redox ions in the electrolyte are controlled by diffusion. On a macroscopic scale, electroneutrality is maintained in the electrode volume by the mutual charge screening of the electrons and holes that are closely packed in the different phases of the nanoporous structure[36,44]. These issues will be discussed in detail in the section relating to various aspects of the nanoporous electrode. At this point, we highlight the fact that the counter movement of the electrons and holes (oxidized ions) across the electrode (Figure 2) increases the probability that they will react with each other. The description of this recombination process in an energy diagram is provided in Figure 3a. Furthermore, the electron traveling across the porous electrode can react with the oxidized dye that was not yet regenerated by the electrolyte. This recombination process is presented in Figure 3b.

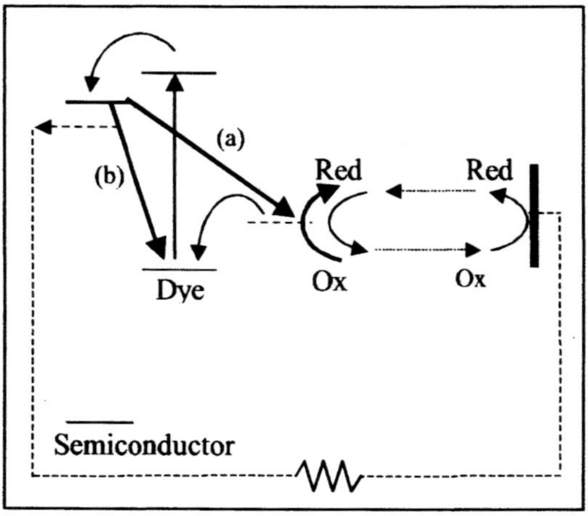

Figure 3. The two major recombination processes of the dye sensitized solar cells.

3.4. The Counter Electrode (Figure 1e)

The electric circuit of the dye sensitized solar cell is closed at the counter electrode. At this side, the oxidized ions are reduced to their original state by the electrons traveling through the external wire. This is the final process that occurs in the cell since it requires the arrival of the oxidized ions to the electrode. The counter electrode utilizes standard electrochemical concepts in order to efficiently reduce the ions[45]. Inefficient reduction leads to a change of the redox ratio and the buildup of potential drops in the system. Furthermore, inefficient operation of the counter electrode increases the hole concentration in the cell. This in turn, increases the rate of the recombination loss process (Figure 3a).

4. MATERIALS

The dye sensitized solar cell consists of a nanoporous electrode, sensitizing dye, hole conducting mediator and a counter electrode. The highest solar energy conversion efficiency was achieved with the type of cell presented schematically in Figure 4[36,41]. The cell was a 10-15 μm thick nanoporous TiO_2 electrode made from particles having a 15-20 nm diameter, an adsorbed monolayer of the N3 (cis-di(isothiocyanato) -N-bis(4,4′-dicarboxy-2,2′-bipyridine) ruthenium(II)) or the Black (tri(cyanato)-2,2'2''-terpyridyl-4,4',4''-tricarboxylate)Ru(II)[30]) dyes, a liquid electrolyte consisting of ca. 0.5M LiI, 0.05M I_2 and 0.2M TBP (4-tert-butylpyridine) in acetonitrile or 3-methoxyproprionitrile, a platinized conducting substrate as counter electrode, and a Baynel, polymer based, sealing. The effort to increase the conversion efficiency of dye sensitized solar cells includes the application of new materials aiming at a high efficiency, stable, solid state cell that will be much less expensive than the alternatives. The literature contains several reviews of the current status of dye sensitized solar cells which discuss available and prospected materials[36,41]. Here we will briefly mention some of these important aspects, emphasizing the nanoporous electrode.

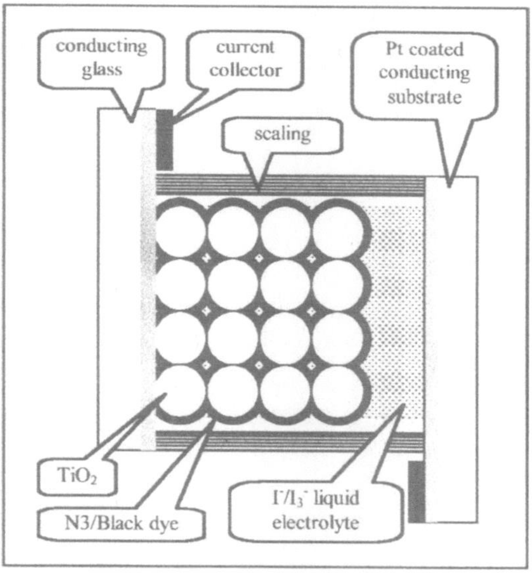

Figure 4. A schematic cross section of a dye sensitized solar cell consisting of a nanoporous electrode, sensitizing dye, hole conducting mediator and a counter electrode.

4.1. Thermodynamic and Kinetic Consideration

Figure 5 presents a diagram of the dye sensitized solar cell, emphasizing the relative energetics in the cell. The cell operation requires the presence of driving forces for two processes: (1) $\Delta E_{injection}$ for the electron injection from the excited level of the dye (E_{D*}) to the conduction band of the semiconductor (E_{CB}) and (2) $\Delta E_{regeneration}$ for the dye regeneration by an electron transfer from the redox electrolyte (E_{RedOx}) to the ground state of the oxidized dye (E_D). The remaining potential ($E_{CB}-E_{RedOx}$) defines the maximum photovoltage that the cell can generate. From the kinetics point of view, efficient cell performance is achieved when these electron transfer processes occur rapidly. Thus when selecting materials for dye sensitized solar cells a fine balance must be achieved between the thermodynamics and kinetics.

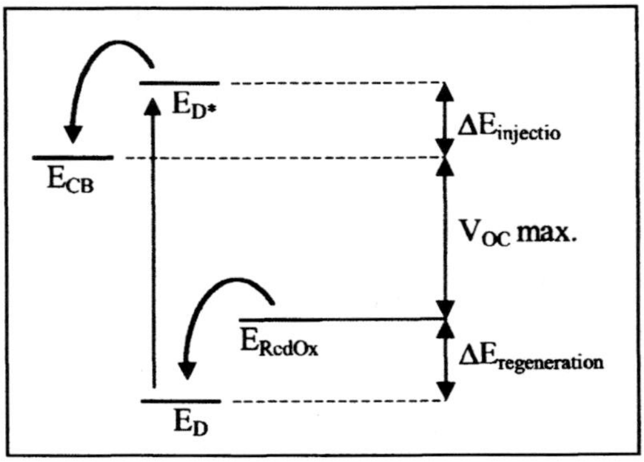

Figure 5. The relative energetics in dye sensitized solar cell that determine the maximum photovoltage and the kinetics of the injection and regeneration processes.

However, Figure 5 presents a simplified picture of the dye sensitized solar cell which is convenient for the basic understanding of the system. As will be discussed later, during the cell operation, the potential of the various cell components can vary with respect to an external reference. Thus, Figure 5 presents, at the best, the relative energetics in the system independent of an external scale[46,47].

4.2. The Nanoporous Electrode

By definition, the dye sensitized solar cells utilize wide band gap semiconductors, while the sensitization to the solar spectrum is performed by the dye. As mentioned above, systems that are based on band gap excitation of nanosize semiconductors exhibit instability in photoelectrochemical systems. Most of the available materials have been tested in dye sensitized solar cells including SnO_2[48-50], rutile TiO_2[51], ZnO[52,53], anatase TiO_2,[7] Nb_2O_5[39,54] and $SrTiO_3$[39]. (The former list was ordered by the conduction band potential, starting with the most positive one). Until now the best performance were achieved with anatase TiO_2.

Despite the extensive investigation of dye sensitized systems a comprehensive understanding with respect to the nanoporous electrode has not yet been achieved. To appreciate this, we compare the performance of TiO_2 and ZnO electrodes. These two materials have similar conduction band potentials and bad gap energies[55]. When fabricated in the same porous geometry, they uptake similar amounts of dye per area (with some restriction regarding ruthenium based dyes[55]). However, the TiO_2-based cells are much more efficient than those containing a ZnO electrode, while the current knowledge cannot fully explain why TiO_2-based cells are more efficient.

The limited choice of materials motivated new approaches in which electrode treatments are used to alter the basic properties of the materials. These treatments include molecular or inorganic modification of the semiconductor surface that shifts the bands to a different potential[56,57]. Composite materials in various configurations were also shown to improve the electrode performance[58-60]. Finally, different porous geometries based on nanorods or mesoporous colloids were tested without any success[61,62].

4.3. The Dye

Many different types of dyes have been tested in dye sensitized solar cells; however, only a few dyes were found to be highly efficient. The leading dyes are the ruthenium complexes, of which the N3 and the Black dye are the most efficient[36,41]. The black dye has a wide absorption spectrum, which expends up to ca. 900 nm, thus providing an excellent match to the solar spectrum. However, since the Black dye is less stable than N3, the latter is usually proffered.

Other types of dyes were tested however these usually resulted in limited performance. Following are some examples. The ultra fast injection rate enables the use of dyes that have a short excited life time such as iron or osmium complexes[38,63]. Molecular dyes such as perylenes, porphyrines and phthalocyanines were tested because of their unique advantages; stability, spectrum, environmental compatibility and price[48,64-69].

Different cell designs related to the dyes have been developed, although to date these attempts have not resulted in efficient cells. One such cell design features the use of two dyes that will match the solar spectrum more efficiently than a single one. Usually dyes having a band at long wavelengths were added to the N3 dye which covers the spectrum up to ca. 650 nm. The failure in this approach is attributed to the low efficiency of the co-adsorbed dye that contributes much less than the N3 per occupied electrode surface[41]. Another approach relates to organic semiconductors. These materials can be applied in a multilayer configuration in which the photoexcited state is transferred by excitons[70-72]. Quantum dots of various materials were also tested but they suffer from extensive instability[2,3,73]. Finally, a tandem cell in which hole injection into the counter electrode provides additional power to the cell was reported[74]. This approach is currently limited by the absence of an efficient nanoporous p-type semiconductor that will accept the holes at the counter electrode.

4.4. The Electrolyte

The electrolyte is currently the weakest point in the application of dye sensitized solar cells to outdoor conditions[41]. It is difficult to seal a cell containing vital solvents for a long operation under extreme environmental conditions. The sealing should prevent both the evaporation of the solvent and the penetration of humidity into the cell. Currently a major effort in dye sensitized solar cells research is directed toward the replacement of the liquid electrolyte by a solid mediator that will transfer the holes from the dye to the counter electrode efficiently. Different materials such as p-type semiconductors[75-78], conducting polymers[41,79] and molten salts[80] were tested but the efficiency achieved with these materials is lower than that of a compatible liquid electrolyte-based cell. This low performance is usually attributed to the conducting performance of the solid mediator and to a pure contact with the dye. However, it is also possible that the lower performance is related to the nanoporous geometry of the cell[44].

At this point, most dye sensitized solar cells utilize the I^-/I_3^- couple in dry solvents such as acetonitrile or 3-methoxyproprionitrile. Other electrolytes decline the cell performance significantly[81,82]. The unique performance of the I^-/I_3^- couple is explained by the negative nature of both ions in the couple that reduces the recombination rate from the negative electrode to the solution, and by a good charge transfer coupling between the dye and this couple[42]. However, attempts to replace this couple are motivated by the anticipated higher photovoltage that could be achieved if a more positive redox electrolyte will be used (see Figure 5)[81,82].

4.5. The Counter Electrode

The counter electrode completes the electrical photogenerated circuit by the re-reduction of the oxidized ions. Efficient reduction requires catalytic properties. The simplest electrode is based on platinum, which by various methods is deposited as a thin layer on a conductive substrate. The platinized electrode can be semitransparent. In addition, other materials such as high surface area carbon seem to operate well. In the case of solid state cells the counter electrode should match the specific mediator.

5. VARIOUS PHOTOELECTROCHEMICAL ASPECTS OF THE NANOPOROUS ELECTRODE

The geometry of the nanoporous electrodes induces special characteristics that differentiate these electrodes from their compact analogues. These porous electrodes are strongly influenced by the following factors: the open structure of the electrodes that permits electrolyte penetration through the entire electrode; the small size of the individual colloidal particles that cannot support a high space charge; and the low inherent conductivity of the semiconductor with respect to the penetrating electrolyte.

In this section, we discuss various aspects of the electrochemistry and photoelectrochemistry of the nanoporous electrodes resulting from their unique characteristics. Although the discussion will related primarily to dye sensitized solar cells it is emphasized that it is also relevant to any electrochemical or photoelectrochemical system that utilizes nanoporous semiconductor electrodes.

5.1. General Considerations

The main benefit in using nanoporous electrodes is their high surface area. The electrode design involves the optimization of two factors: the surface area per electrode volume and the pore size. Both factors are controlled primarily by the particle size i.e., decreasing the particle diameter results in both an increase of the surface area per electrode volume, and a decrease of the pore average diameter. The surface area per electrode volume is important because it defines the electrode thickness for a desirable surface area. We provide below some characteristics of nanoporous electrodes that are thickness dependent. For example, an applied bias is unequally distributed across the electrode and electrons diffuse across the electrode only for a limited distance. Thus, it is always more efficient to use a relatively thin electrode. On the other hand one has to keep the pores large enough to allow sufficient electrolyte diffusion. Both parameters are highly dependent on the specific application in which the electrodes are used. In dye sensitized solar cells for example, the electrodes do not exceed 15 μm with particles of 15-25 nm.

5.2. Electric Potential Distribution and Charge Screening

Various electrochemical measurements of the nanoporous electrodes and theoretical calculations show that applied potentials are unequally distributed throughout the semiconductor electrode-electrolyte interface[44,83-85]. For moderate applied bias, the effective potential is equal to the applied potential only in the conducting substrate and in the few semiconductor layers next to it. The effect of the applied bias decreases as the distance from the substrate increases, because ion motion through the film neutralizes the applied field over a short range[44,85,86].

The unequal potential distribution can be explained by the semiconducting nature of the electrode and the porous geometry of the films. Figure 6 shows a simplified schematic view of the interconnected colloids represented as cross-linked columns. Electrolyte solution is able to penetrate through the semiconductor film (columns) thus creating, under an applied potential, a potential distribution unlike that in a compact electrode.

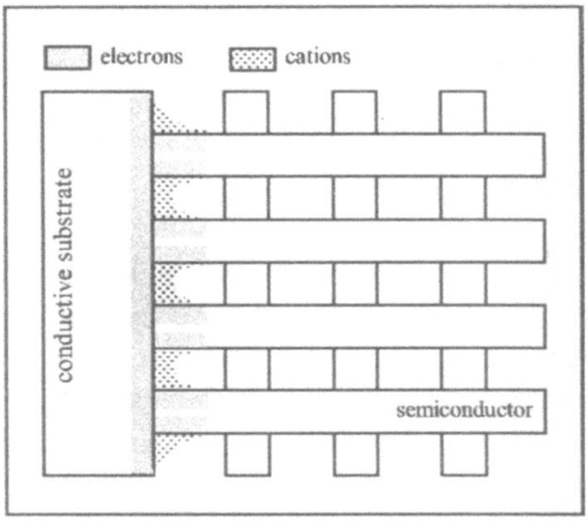

Figure 6. The unequal applied potential distribution in the interconnected colloids represented as cross-linked columns.

A schematic representation of this unusual field distribution for one column is presented in Figure 7. Figures 7 a-b show, respectively, the approximate electric field force lines and the equipotential curves generated by a negative potential applied to such a representative semiconducting column immersed in electrolyte solution. Figure 7c shows the expected equipotential curves for a similar metallic or highly conducting column. In the latter case, there can be no potential drop across the metal, thus the entire applied potential drops at the metal-solution interface. In contrast, when the porous film is insulating or semiconducting (Figure 7a-b), the ability of the electrolyte ions to move toward the conducting substrate forces most of the potential to drop near the substrate. This implies that at increasing distance from the substrate, the semiconductor-solution potential difference decreases and, correspondingly, that the concentration of the electrons in the semiconductor decreases. It is important to note that the difference in potential distribution between the metallic and the semiconducting cases results from their different conductivities relative to solution. In fact, this parameter determines the exact potential distribution for any given film geometry.

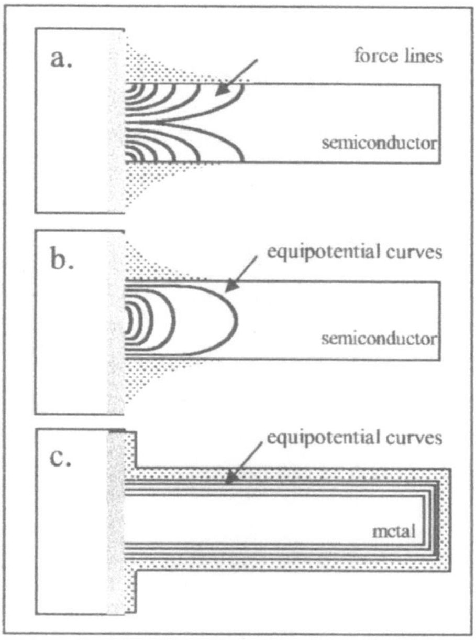

Figure 7. A schematic representation of the unusual field distribution for one column of Figure 6. (a) the equipotential curves and (b) the approximate electric field force lines generated by a negative potential applied to a semiconducting column. (c) the corresponding electric field force lines in a conducting column.

Most nanoporous semiconductor electrodes consist of weakly doped ($\sim 10^{16}$ cm^{-3}) colloids[87], and are therefore expected to approximate the semiconducting behavior illustrated in Figure 7b. Electrons injected by the applied potential can, however, significantly increase the semiconductor conductivity. Since the effect of the applied negative potential, and thus the electron concentration, decreases across the nanoporous film as the distance from the conductive substrate increases (Figure 7b), a corresponding gradual change in the film's conductivity can be expected. For n-type semiconductors, under negative bias, the semiconductor nearest the substrate should become electrochemically active, while the outer part is still effectively insulating. The border between the two zones is anticipated to move away from the conductive substrate as the applied potential becomes more negative and a larger portion of the semiconductor film becomes conductive. In other words, an increase in the electrochemically active surface area is expected when the applied bias is scanned negatively.

An intuitive and unambiguous measure of the potential distribution in nanoporous semiconductor electrodes was provided by electrochemical dye desorption experiments. Dyes are desorbed from the semiconductor film above a threshold negative potential[49], providing a means to directly visualize the spatial variation in the potential distribution by absorbance measurements. If the applied negative potential decreases across the film as proposed above, the dye should desorb only near the conductive substrate under moderate negative bias. Dye desorption measurements performed with various semiconductor materials including TiO_2, SnO_2, ZnO and Nb_2O_5 indeed show this behaviour[44,49,59].

A special electrode is needed to perform absorbance measurements as a function of the distance from the electrical contact; a porous semiconductor film deposited on a nonconductive glass with a metal contact that is evaporated at one end of the film[44]. After dye adsorption, the electrode is subjected to a constant electrochemical potential that is sufficient to desorb the dye. After the electrochemical desorption, the transmission along the electrode is measured with a laser spot to detect the dye concentration as a function of the distance from the metal contact.

5.3. The Conductive Substrate

The conductive substrate serves as a current collector of the porous semiconductor film. However it is also exposed to the electrolyte solution in areas that are not covered with the semiconductor particles. Thus, when using a nanoporous electrode one must account for the direct substrate-electrolyte contact, unless it is electrochemically passivated. The exposed substrate can participate in the electrode operation and affect the results. The substrate contribution to the electrode process should be minor when the interaction of the substrate and the film with the electrolyte exhibit similar kinetics because of the huge difference in surface area between them. However in cases where the substrate is more active, or when the semiconductor film is partially passivated, the substrate contribution may be significant, thus altering the interpretation of the results[44,83-85].

Various methods for the passivation of the exposed substrate were reported. The passivation material should not necessarily be similar to that of the porous film. Usually the passivation is achieved by the formation of a compact semiconductor layer prior to the particles deposition or by coating the entire electrode after its formation[41,88]. Useful blocking can also be achieved by electrochemical deposition of isolating polymers at potentials in which the semiconductor is inactive thus ensuring the specific coating[89].

5.4. Band Bending

The individual particles in the nanoporous electrode are small and usually weakly doped. Thus, there is no space in the particles to allow band bending at the electrode-electrolyte interface. This characteristic implies that external perturbations may cause the bands to shift unlike the band bending that occurs in compact electrodes[22]. In addition, electrons are not drifted away from the electrode surface by the space layer field in the absence of a space charge layer. Thus an interpretation of the results or a system design, which are related to nanoporous electrodes, should account for these effects.

The absence of a space charge layer has been discussed extensively in electrochemical impedance studies of nanoporous electrodes[44,85,90,91]. The space charge dominates the impedance spectra of compact electrodes thus being the basis of the Mott Schottky analysis. In contrast, in the case of nanoporous electrodes, the impedance measures the Helmholtz capacitance in addition to transport phenomena. Consequently, the Mott Schottky analysis is irrelevant to nanoporous electrodes, and other methods such as spectroelectrochemical measurements, are used to determine the conduction band potential[92-94].

5.5. Core Shell Electrodes

In sections 3.3 we discussed the recombination processes of dye sensitized solar cells. In brief, during the operation of a dye sensitized solar cell, the injected electrons diffuse through the TiO_2 film towards the conducting substrate, while the oxidized ions move in the opposite direction to be regenerated at the counter electrode. The porous geometry that permits electrolyte presence through the entire electrode provides a high surface area for the recombination of the photoinjected electrons and the holes in the dye layer, or in the electrolyte (ions)[95-98]. The small size of the individual colloidal particles in the nanoporous electrode cannot support a high space charge[85,86,95]. Thus, in the absence of band bending at the semiconductor surface, there is no energy barrier that slows the recombination process, as usually occurs in bulk electrodes. Therefore, slowing the recombination rate is a major task in attempts to increase the efficiency of Dye sensitized solar cells.

Most of the effort to improve the efficiency of Dye sensitized solar cells by suppression of the recombination process involves two basic approaches[99,100]. The first approach physically blocks the electrode area that is not covered with dye. The second approach involves the formation of an energy gradient that directs the electrons towards the substrate. The physical blocking involves adsorption of insulating molecules or polymerization of an insulating layer on the semiconductor surface after the dye adsorption[49,89]. This blocking approach faces a complexity related of mutual effects between the insulating layer and the dye. Furthermore, this approach requires a complex process that will ensure that the coating will not separate the dye from the solution. On the other hand, the energy gradient approach involves composite material nanoporous electrodes in which the two materials differ by their conduction band potential[58,99,101]. Arranging these materials in the correct geometry is expected to drive the electrons to the desired direction by energy considerations i.e., the electrons will favor the material having the more positive conduction band. This gradient approach requires a very specific electrode design which ensures that the electrons will not encounter energy barriers (the more negative material) while diffusing to the current collector. As indicated in Figure 8a-b, this is not the case when the electrode is made from core shell colloids or from a simple mixture of two semiconductor particles.

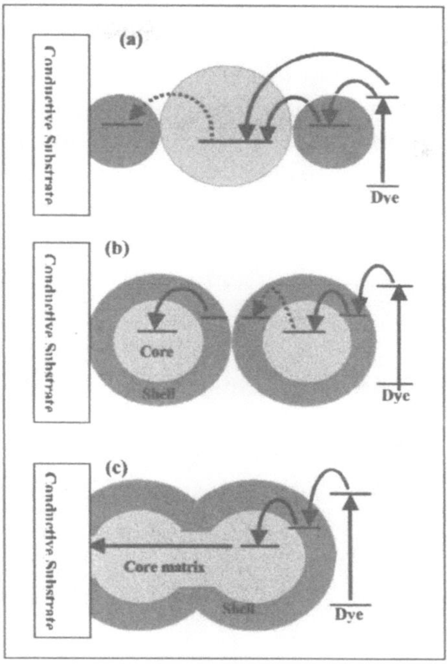

Figure 8. Various energy gradient approaches for efficient electrode design. The materials composing the electrodes differ by their conduction band potential. An inappropriate composition (a and b) can induce energy barriers into the electrode.

Recently, a new type of composite material nanoporous electrode, which applies the energy gradient approach, was reported[58]. This core-shell electrode was made by coating a nanoporous TiO_2 matrix with an Nb_2O_5 layer. The conduction band potential of the Nb_2O_5 is approximately 100 mV negative of the potential of the TiO_2[102]. As illustrated in Figure 8c this potential difference forms an energy barrier at the electrode-electrolyte interface. Thus, electrons injected into the electrode are driven away from the electrode surface into the TiO_2 core which slows the recombination rate and increases the cell's efficiency (Figure 8c). In other words, the coating forms an inherent energy barrier at the electrode surface. A comparison of two similar Dye sensitized solar cells that differ only by their nanoporous electrodes, shows that the new core shell electrode is superior to the standard one concerning all cell parameters thus increasing the conversion efficiency significantly[58-60].

The superiority of the core shell electrode is not limited to its operation in dye sensitized solar cells. This electrode also improves the distribution of applied potential across the nanoporous electrode which is important to other electrochemical and photoelectrochemical applications (see section 5.2). The unequal potential distribution depends on the ratio between the conductivity of the semiconductor and the electrolyte that penetrates the pores. When an energy barrier is formed at the surface of the porous electrode, it is possible to accumulate more electrons in the semiconductor thus increasing its conductivity with respect to a noncoated analogue. The increased conductivity of the semiconductor drives the potential distribution towards better uniformity.

5.6. Electron Motion in the Nanoporous Electrode

The mechanism by which electrons move through the semiconductor film is an important factor in systems that utilize the nanoporous electrodes. Models describing the electron motion by diffusion have been developed in some studies based on various measuring techniques[103-105]. The diffusion motion occurs in both photoelectrochemical and electrochemical systems[44,85].

The diffusion model may be rationalized by the short range screening described above (section 5.2). Because of the porous nature of the electrode, ions can migrate through the film to neutralize any electric fields (including those caused by the moving electrons) over very short distances. Therefore, under normal operating conditions, there should be essentially no macroscopic electric fields in the TiO_2 film, depending on the electrolyte concentration and composition[44,45,85]. Accordingly, during steady state illumination of a dye sensitized solar cell, the injected electrons experience little or no electric field, so their motion is governed primarily by concentration gradients, that is, diffusion.

The diffusional motion of the electrons in the semiconductor film was found to be trap-limited[106-111]. Electronic states that are located in the band gap, trap and release electrons, thus significantly slowing the diffusion rate. Filling the traps by photo or electro-injected electrons decreases the trapping depth and thus the trapping time shortens. In other words, the electron mobility in the semiconductor film changes as a function of the electron concentration in the film[106-112].

5.7. Relative Energetics in the Nanoporous System

One of the most fundamental properties of dye sensitized solar cells is the relative energies at the semiconductor/dye /electrolyte interface. In a dye sensitized solar cell, the dye excited state potential has to be more negative than the semiconductor conduction band potential to enable the electron injection, and the oxidation potential of the dye must be more positive than the redox couple in the electrolyte solution to provide the driving force for the hole transfer[26,88]. Within these limits, the cell performance is affected by the exact position of the relative potentials. A change in either of the two driving forces impacts both the short circuit photocurrent and the open circuit photovoltage of the cell[88,113]. Therefore, optimization of the dye sensitized solar cells must involve consideration of the semiconductor, dye and electrolyte energetics[26].

The following four potentials that affect the semiconductor/ dye/electrolyte interface can be measured separately: (1) the conduction band potential of the high surface area semiconductor can be resolved by measurements of its absorption[92] or reflection[114] during a potential sweep; (2) the oxidation potential of the dye in solution and (3) the potential of the redox couple can be measured by cyclic voltammetry[115]; and (4) the dye's excited state potential can be estimated from its oxidation potential and the excited state energy[48,88]. However, measurements of these potentials under working cell conditions, i.e., when the dye is adsorbed onto the semiconductor and the electrolyte solution is present, are more complicated. The complexity arises primarily from the limited electrochemical window resulting from the insulating nature of the TiO_2 at positive bias[47] and the tendency of the dye to desorb at negative bias[44]. It is, therefore, usually assumed that the potentials measured individually still pertain when the complete solar cell is assembled.

Various measurements of the adsorbed dye potential indicate that it is affected by its surrounding i.e. the semiconductor and the electrolyte. The redox potential of the dye was investigated by chemical oxidation, spectro-electrochemical reduction and potential dependent photoluminescence[24,46,47]. The chemical oxidation potential measurements are based on the well-known movement of the TiO_2 conduction band by 59 mV per pH unit[24,116]. The dyes exhibit a pH dependent redox potential induced by adsorption to TiO_2 or to insulating Al_2O_3, although the potential of the dye in solution is independent of pH. The magnitude of the pH sensitivity varies between 21 and 53 mV per pH unit depending on the dye structure. Electrochemical experiments show that the redox potential of the dye is also affected by an externally applied bias. This sensitivity to pH and electrical bias depends both on the ability of electrolyte ions to penetrate the dye layer and on the dye structure. The crucial parameter is the position of the specifically adsorbed dye relative to the ionic double layer (Helmholtz and diffuse layers). The adsorption induced potential changes were observed for several ruthenium based dyes, phthalocyanine dyes and perylene dyes. These changes are therefore considered as a general characteristic of the dye sensitized system.

The movement of the dye redox potential is attributed to the presence of an electric field across the dye layer. Being inside the electric field, the redox center of the dye is affected so that its measured potential shifts with the substrate. Note that this effect does not require the existence of a compact monolayer, but only that some fraction of the electric field drops across the dye. The dye is therefore at least partially inside the ionic double layer at the substrate/solution interface. Indeed, it was recently found that a field caused by co-adsorption of a dipole also shifts the dye potential[57].

5.8. The Open Circuit Potential in Dye Sensitized Solar Cells

The photovoltage generated in the dye sensitized solar cells is measured at the two external contacts. This leaves some uncertainty regarding the location and the nature of this photo generated potential drop in the cell. Various studies locate the potential drop at either the semiconductor/electrolyte interface or the semiconductor/substrate interface while providing different photovoltage models. In practice, a commonly accepted description of the dependence of the open circuit voltage (V_{OC}) on the light intensity (I_0) is not achieved. Here we overview a very simple picture of the photovoltage dependence on the light intensity. This picture provides, so to speak an "ideal" model of the dye sensitized solar cell, but it is known to fail quantitatively. Thus this simple model should be regarded only as a starting point, needed in order to understand the more realistic mechanisms of V_{OC} that are currently suggested. A short overview and literature references will be provided below.

In the simple model we first analyze the relationship between the photovoltage and the free electron density in the semiconductor[42],

$$n = n_0 \exp\left(eV_{oc} / kT\right) \tag{1}$$

where n_0 is the value at dark equilibrium, and e and k represent the positive elementary charge and the Boltzmann constant, respectively.

A justification of eq. 1 follows. In the dark, the location of the Fermi level of electrons in the semiconductor is determined by the redox electrolyte, so that the semiconductor is probably in depletion, with an electron density given by

$$n_0 = N_c \exp[-(E_{c0} - E_{F0})/kT] \tag{2}$$

where N_c is the density of states in the conduction band, E_c is the energy at the bottom of the conduction band, and E_F is the energy of the Fermi level, and the subscript 0 denotes the value at dark equilibrium for each quantity.

When the cell is irradiated with photons so that the dye injects electrons, at open circuit, the electron density increases throughout the semiconductor, as described by a more positive value of the Fermi level:

$$n = N_c \exp[-(E_{c0} - E_F)/kT] \tag{3}$$

In this simple approach, it is assumed that the conduction band edge stays fixed at its dark equilibrium value; this is not a necessary assumption but it is the simplest and has been widely adopted. The variation of the Fermi level in the semiconductor gives rise to the measured photovoltage V_{OC}:

$$eV_{oc} = E_F - E_{F0} \tag{4}$$

Combining eqs. 2-4 one gets eq. 1.

Secondly, in the most commonly accepted view, the free electron density at a given light intensity I_0 is described by a balance over time between a homogeneous generation rate $G = \alpha I_0$ per unit volume (where α is the effective absorption coefficient) and the rate at which the electrons are lost from the semiconductor, as follows:

$$\frac{\partial n}{\partial t} = G - \frac{n - n_0}{\tau_r} \tag{5}$$

where τ_r is an average lifetime of electrons in the semiconductor, in stationary conditions, is

$$n = \alpha \tau_r I_0 \tag{6}$$

provided that $n >> n_0$.

The relationship between photovoltage and light intensity are provided by eqs. 1 and 6 as follows:

$$V_{oc} = \frac{1}{p}\left[\log\left(\frac{\alpha\tau_r}{n_0}\right) + \log I_0\right] \tag{7}$$

where

$$p = \frac{e}{2.30kT} \tag{8}$$

In this simple model, a description of photovoltage versus light intensity in the dye sensitized solar cell is provided by two parameters: n_0, which is determined by the initial position of the conduction band and therefore reflects mainly the thermodynamics, and τ_r, which reflects the kinetics.

According to eq. 7, the curve I_0 vs. V_{OC} should be a straight line with a slope of $p = (59\text{mV})^{-1} = 17\text{V}^{-1}$, at room temperature. Experimentally, it is generally found that the plot log I_0 versus. V_{OC} follows approximately a straight line but the slope is in the order of 0.5-0.7 x 17V^{-1} [42,112].

Several approaches were adopted in order to account for the measured correlation of I_0 versus V_{OC}. In one approach, the kinetic part of eq. 7 (τ_r) was modified based on time and frequency resolved measurements that predicted a recombination order higher than 1 [42,117,118]. This approach further involves the electron concentration dependent mobility, often described as an increase of the mobility with the increase of electron concentration by a power law. A second approach assigns the measured I_0 (V_{OC}) slope to a shift of the semiconductor bands during the increase of electron concentration in the semiconductor. In this approach the thermodynamic part of eq. 7 (n_0) is used; based on steady state simultaneous measurement of the V_{OC} and the semiconductor resistance. The semiconductor band shift may be described as Fermi level pinning[112]. Finally, a third approach assigns the potential drop to the semiconductor/substrate interface[105]. In this view the electrons travel with a counter ion to the substrate. Here they are separated as the electron enters the conductive substrate leaving the ion in solution. Comprehensive models for the three approaches can be found in the original publications quoted above.

6. REFERENCES

(1) Athanassov, Y.; Rotzinger, F. P.; Pechy, P.; Gratzel, M. *J. Phys. Chem. B* **1997**, *101*, 2558.
(2) Zaban, A.; Micic, O.; Gregg, B. A.; Nozik, A. J. *Langmuir* **1998**, *14*, 3153-3156.
(3) Hoyer, P.; Konenkamp, R. *Appl. Phys. Lett.* **1995**, *66*, 349.
(4) Gerfin, T.; Gratzel, M.; Walder, L. Molecular and Supermolecular Surface Modification of Nanocrystalline TiO2 Films: Charge Separating and Charge Injecting Devices. In *Molecular Level Artificial Photosynthetic Materials*; Karlin, K. D., Ed.; John Wiley & Sons, Inc.: New York, 1997; Vol. 44; pp 345-393.
(5) Gregg, B. G. *Endeavour* **1997**, *21*, 52-55.
(6) Bechinger, C.; Ferrere, S.; Zaban, A.; Sprague, J.; Gregg, B. A. *Nature* **1996**, *383*, 608-610.
(7) O'Regan, B.; Gratzel, M. *Nature* **1991**, *353*, 737-740.
(8) Hagfeldt, A.; Didriksson, B.; Palmqvist, T.; Lindstrom, H.; Sodergren, S.; Rensmo, H.; Lindquist, S. E. *Sol. Energy Mater. Sol. Cells* **1994**, *31*, 481.
(9) Nozik, A. J.; Memming, R. *J. Phys. Chem.* **1996**, *100*, 13061.
(10) Micic, O. I.; Cheong, H. M.; Fu, H.; Zunger, A.; Sprague, J. R.; Mascarenhas, A.; Nozik, A. J. *J. Phys. Chem. B* **1997**, *101*, 4904.
(11) Herhold, A. B.; Tolbert, S. H.; Guzelian, A. A.; Alivisatos, A. P. A comparison of pressure induced structural transformations in CdSe, InP and Si nanocrystals. In *fine particles science and technology*; Pelizzetti, E., Ed.; Kluwer Academic Publishers: London, 1996; pp 331-342.
(12) Chen, C. C.; Herhold, A. B.; Johnson, C. S.; Alivisatos, A. P. *Science* **1997**, *276*, 398.
(13) Gribb, A. A.; Banfield, J. F. *Am. Mineral.* **1997**, *82*, 717.
(14) Zhang, H. Z.; Banfield, J. F. *J. Mater. Chem.* **1998**, *8*, 2073.
(15) Zhang, H. Z.; F., B. J. *Am. Mineral.* **1999**, *84*, 528-535.
(16) Zaban, A.; Aruna, S. T.; Tirosh, S.; Gregg, B. A.; Mastai, Y. *J. Phys. Chem. B* **2000**, *104*, 4130.
(17) Aruna, S. T.; Tirosh, S.; Zaban, A. *J. Mater. Chem.* **2000**, *10*, 2388-2391.
(18) Liu, Y.; Claus, R. O. *J. Am. Chem. Soc.* **1997**, *119*, 5273.
(19) Vinodgopal, K.; Bedja, I.; Kamat, P. V. *Chem. Mater.* **1996**, *8*, 2180.
(20) Kamat, P. V. Native and Surface Modified Semiconductor Nanoclusters. In *Molecular Level Artificial Photosynthetic Materials*; Karlin, K. D., Ed.; John Wiley & Sons, Inc.: New York, 1997; Vol. 44; pp 273-343.
(21) Torimoto, T.; Fox, R. J.; III; Fox, M. A. *J. Electrochem. Soc.* **1996**, *143*, 3712.
(22) Kavan, L.; Gratzel, M.; Gilbert, S. E.; Klemenz, C.; Scheel, H. J. *J. Am. Chem. Soc.* **1996**, *118*, 6716.
(23) Meyer, G. J. *J. Chem. Educ.* **1997**, *74*, 652.
(24) Gerischer, H. *Photochem. Photobiol.* **1972**, *16*, 243.
(25) Memming, R. *Photochem. Photobiol.* **1972**, *16*, 325.
(26) Parkinson, B. A.; Spitler, M. T. *Electrochim. Acta.* **1992**, *37*, 943.
(27) Argazzi, R.; Bignozzi, C. A.; Heimer, T. A.; Meyer, G. J. *Inorg. Chem.* **1997**, *36*, 2-3.
(28) Bonhote, P.; Moser, J. E.; Vlachopoulos, N.; Walder, L.; Zakiruddin, S. M.; Humphry-Baker, R.; Pechy, P.; Gratzel, M. *Chem. Commun.* **1996**, 10.
(29) Gratzel, M. "Photoelectrochemical Solar Energy Conversion by Dye Sensitization"; First Conference on Future Generation Photovoltaic Technologies, 1997, Denver CO.
(30) Nazeeruddin, M. K.; Pechy, P.; Gratzel, M. *Chem. Commun.* **1997**, 1705.

(31) Ellingson, R. J.; Asbury, J. B.; Ferrere, S.; Ghosh, H. N.; Sprague, J. R.; Lian, T. Q.; Nozik, A. J. *J. Phys. Chem. B* **1998**, *102*, 6455.

(32) Ghosh, H. N.; Asbury, J. B.; Lian, T. Q. *J. Phys. Chem. B* **1998**, *102*, 6482.

(33) Ghosh, H. N.; Asbury, J. B.; Weng, Y. X.; Lian, T. Q. *J. Phys. Chem. B* **1998**, *102*, 10208.

(34) Burfeint, B.; Hannappel, T.; Storck, W.; Willig, F. *J. Phys. Chem.* **1996**, *100*, 16463-16465.

(35) Rehm, J.; McLendon, G. L.; Nagasawa, Y.; Yoshihara, K.; Moser, J.; Gratzel, M. *J. Phys. Chem.* **1996**, *100*, 9577.

(36) Hagfeldt, A.; Gratzel, M. *Acc. Chem. Res.* **2000**, *33*, 269.

(37) Asbury, J. B.; Hao, E.; Wang, Y. Q.; Ghosh, H. N.; Lian, T. Q. *J Phys Chem B* **2001**, *105*, 4545.

(38) Ferrere, S.; Gregg, B. A. *J. Amer. Chem. Soc.* **1998**, *120*, 843.

(39) Lenzmann, F.; Krueger, J.; Burnside, S.; Brooks, K.; Gratzel, M.; Gal, D.; Ruhle, S.; Cahen, D. *J. Phys. Chem. B* **2001**, *105*, 6347.

(40) Remakrishna, G.; Gohosh, H. *J. Phys. Chem. B* **2001**, In Press.

(41) Gratzel, M. *Prog. Photovoltaics* **2000**, *8*, 171.

(42) Huang, S. Y.; Schlichthorl, G.; Nozik, A. J.; Gratzel, M.; Frank, A. J. *J. Phys. Chem. B* **1997**, *101*, 2576-2582.

(43) Park, N. G.; Schlichthorl, G.; van de Lagemaat, J.; Cheong, H. M.; Mascarenhas, A.; Frank, A. J. *J. Phys. Chem.* **1999**, *103*, 3308-3314.

(44) Zaban, A.; Meier, A.; Gregg, B. A. *J. Phys. Chem. B* **1997**, *101*, 7985.

(45) Bard, A. J.; Faulkner, L. R. *Electrochemical Methods Fundamental and Applications*; John Wiley & Sons: New York, 1980.

(46) Zaban, A.; Ferrere, S.; Gregg, B. A. *J. Phys. Chem. B* **1998**, *102*, 452-460.

(47) Zaban, A.; Ferrere, S.; Sprague, J.; Gregg, B. A. *J. Phys. Chem.* **1997**, *101*, 55-57.

(48) Ferrere, S.; Zaban, A.; Gregg, B. A. *J. Phys. Chem. B* **1997**, *101*, 4490-4493.

(49) Kamat, P. V.; Bedja, I.; Hotchandani, S.; Patterson, L. K. *J. Phys. Chem.* **1996**, *100*, 4900 - 4908.

(50) Chappel, S.; Zaban, A. *Solar Energy Mater. Solar Cells* **2000**, In Press.

(51) Park, N. G.; van de Lagemaat, J.; Frank, A. J. *J. Phys. Chem. B* **2000**, *104*, 8989.

(52) Hoyer, P.; Weller, H. *J. Phys. Chem.* **1995**, *99*, 14096-14100.

(53) Rensmo, H.; Keis, K.; Lindstrom, H.; Sodergren, S.; Solbrand, A.; Hagfeldt, A.; Lindquist, S. E. *J. Phys. Chem. B* **1997**, *101*, 2598.

(54) Sayama, K.; Sugihara, H.; Arakawa, H. *Chem. Mater.* **1998**, *10*, 3825-3832.

(55) Keis, K. Nanostructured ZnO Electrodes for Solar Cell Applications. Ph.D., Uppsala University, 2001.

(56) Kruger, J.; Bach, U.; Gratzel, M. *Adv. Mater.* **2000**, *12*, 447.

(57) Zaban, A.; Chen, S. G.; Pizem, H.; Sukenik, H. S. "Molecular Modification of the Electronic Properties of Nanoporous TiO2 electrodes"; The American Chemical Soc., 2001, Chicago, Illinois.

(58) Zaban, A.; Chen, S. G.; Chappel, S.; Gregg, B. A. *Chem. Comm.* **2000**, *22*, 2231-2232.

(59) Chen, S. G.; Chappel, S.; Diamant, Y.; Zaban, A. *Chem. Mater.* **2001**, In Press.

(60) Chappel, S.; Chen, S. G.; Zaban, A. **2001**, submitted for publication.

(61) Chen, S. G.; Vijaya Kumar, R.; Gedanken, A.; Zaban, A. *Israel J. Chem.* **2001**, In Press.

(62) Wang, Y. Q.; Chen, S. G.; Tang, X. H.; Palchik, O.; Zaban, A.; Koltypin, Y.; Gedanken, A. *J. Mater. Chem.* **2000**, In Press.

(63) Kuciauskas, D.; Sauve, G.; Freund, M. S.; Gray, H. B.; Winkler, D. R.; Lewis, N. S. "Nanocrystalline titanium dioxide solar cells sensitized with ruthenium or osmium

polypyridyl complexes: Photoelectrochemical studies and electron transfer dynamics"; The American Chemical Soc., 2001, Chicago, Illinois.

(64) Sayama, K.; Hara, K.; Mori, N.; Satsuki, M.; Suga, S.; Tsukagoshi, S.; Abe, Y.; Sugihara, H.; Arakawa, H. *Chem. Comm.* **2000**, 1173.

(65) Deng, H. H.; Lu, Z. H.; Mao, H. F.; Xu, H. J. *Chemical Physics* **1997**, *221*, 323-331.

(66) Deng, H. H.; Mao, H. F.; Lu, Z. H.; Li, J. M.; Xu, H. J. *Journal of Photochemistry and Photobiology A Chemistry* **1997**, *110*, 47-52.

(67) Fang, J. H.; Wu, J. W.; Lu, X. M.; Gu, J. H.; Lu, Z. H. *Journal of Vacuum Science & Technology B* **1997**, *15*, 1468-1470.

(68) Boschloo, G. K.; Goossens, A. *J. Phys. Chem.* **1996**, *100*, 19489.

(69) Kay, A.; Gratzel, M. *J. Phys. Chem.* **1993**, *97*, 6272-6277.

(70) Diamant, Y.; Zaban, A. *J. Electrochem. Soc.* **2001**, In Press.

(71) Zaban, A.; Diamant, Y. *J. Phys. Chem. B* **2000**, *104*, 10043-10046.

(72) Gregg, B. A. *Chem. Phys. Lett.* **1996**, *258*, 376.

(73) Nasr, C.; Kamat, P. V.; Hotchandani, S. *J. Electroanal. Chem.* **1997**, *420*, 201.

(74) He, J. J.; Lindstrom, H.; Hagfeldt, A.; Lindquist, S. E. *J. Phys. Chem. B* **1999**, *103*, 8940.

(75) O'Regan, B.; Schwartz, D. T. *J. Appl. Phys.* **1996**, *80*, 4749.

(76) O'Regan, B.; Schwartz, D. T. "Efficient Photo- Hole Injection from Cyanine Dyes into Electrodeposite CuSCN: A prospective p-Type Material for a Dye Sensitized p-n Hetrojunction"; Electrochem. Soc. Meeting, 1995.

(77) Tennakone, K.; Kumara, G. R. R. A.; Kumarasinghe, A. R.; Wijayantha, K. G. U.; Sirimanne, P. M. *Semicond. Sci. Technol.* **1995**, *10*, 1689.

(78) Tennakone, K.; Kumara, G. R. R. A.; Wijayantha, K. G. U.; Kottegoda, I. R. M.; Perera, V. P. S.; Aponsu, G. M. L. P. *Journal of Photochemistry and Photobiology A Chemistry* **1997**, *108*, 175-177.

(79) Nogueira, A. F.; Durrant, J. R.; De Paoli, M. A. *Adv. Mater.* **2001**, *13*, 826.

(80) Papageorgiou, N.; Athanassov, Y.; Armond, M.; Bonhote, P.; Petterson, H.; Azom, A.; Gratzel, M. *J. Electrochem. Soc.* **1996**, *143*, 3099.

(81) Pichot, F.; Gregg, B. A. *J. Phys. Chem. B* **2000**, *104*, 6.

(82) Oskam, G.; Bergeron, B. V.; Meyer, G. J.; Searson, P. C. *J. Phys. Chem. B* **2001**, *105*, 6867.

(83) Fabregat-Santiago, F.; Garcia-Belmonte, G.; Bisquert, J.; Zaban, A. **2001**, submitted for publication.

(84) Fabregat-Santiago, F.; Garcia-Belmonte, G.; Bisquert, J.; Zaban, A.; Salvador, P. **2001**, submitted for publication.

(85) Bisquert, J.; Garcia-Belmonte, G.; Fabregat-Santiago, F. *J. Solid State Electrochem.* **1999**, *3*, 337.

(86) Cahen, D.; Hodes, G.; Gratzel, M.; Guillemoles, J. F.; Riess, I. *J. Phys. Chem. B* **2000**, *104*, 2053.

(87) O'Regan, B.; Moser, J.; Anderson, M.; Gratzel, M. *J. Phys. Chem.* **1990**, *94*, 8720.

(88) Hagfeldt, A.; Gratzel, M. *Chem. Rev.* **1995**, *95*, 49-68.

(89) Gregg, B. A.; Pichot, F.; Ferrere, S.; Fields, C. L. *J. Phys. Chem. B* **2001**, *105*, 1422.

(90) Wahl, A.; Augustynski, J. *J. Phys. Chem. B* **1998**, *102*, 7820.

(91) Kavan, L.; Gratzel, M.; Rathousky, J.; Zukal, A. *J. Electrochem. Soc.* **1996**, *143*, 394.

(92) Rothenberger, G.; Fitzmaurice, D.; Gratzel, M. *J. Phys. Chem.* **1992**, *96*, 5983-5986.

(93) Boschloo, G.; Fitzmaurice, D. *J.Phys. Chem. B* **1999**, *103*, 2228.

(94) Boschloo, G.; Fitzmaurice, D. *J.Phys. Chem. B* **1999**, *103*, 3093.

(95) Hagfeldt, A.; Lindquist, S. E.; Gratzel, M. *Sol. Energy Mater. Sol. Cells* **1994**, *32*, 245.

(96) Oskam, G.; Bergeron, B. V.; Meyer, G. J.; Searson, P. C. *J. Phys. Chem. B* **2001**, *105*, 6867.

(97) Tachibana, Y.; Moser, J. E.; Gratzel, M.; Klug, D.; Durrant, J. R. *J. Phys. Chem.* **1996**, *100*, 20056.

(98) Nelson, J.; Haque, S. A.; Klug, D. R.; Durrant, J. R. *Phys Rev B* **2001**, *6320*, 5321.

(99) Tada, H.; Hattori, A. *J. Phys. Chem. B* **2000**, *104*, 4585.

(100) Bedja, I.; Hotchandani, S.; Kamat, P. V. *J Phys Chem* **1994**, *98*, 4133.

(101) Bedja, I.; Kamat, P. V. *J Phys Chem* **1995**, *99*, 9182.

(102) Hodes, G.; Howell, I. D. J.; Peter, L. M. *J. Electrochem. Soc.* **1992**, *139*, 3136.

(103) Cao, F.; Oskam, G.; Meyer, G. J.; Searson, P. C. *J. Phys. Chem.* **1996**, *100*, 17021.

(104) Boschloo, G. K.; Goossens, A.; Schoonman, J. *J. Electroanal. Chem.* **1997**, *428*, 25-32.

(105) Schwarzburg, K.; Willig, F. *J. Phys. Chem. B* **1999**, *103*, 5743 -5746.

(106) Meulenkamp, E. A. *J. Phys. Chem. B* **1999**, *103*, 7831.

(107) de Jongh, P. E.; Vanmaekelbergh, D. *J. Phys. Chem. B* **1997**, *101*, 2716-2722.

(108) Konenkamp, R. *Phys. Rev. B* **2000**, *61*, 11057.

(109) Dloczik, L.; Ileperuma, O.; Lauerman, I.; Peter, L. M.; Ponomarev, E. A.; Redmond, G.; Shaw, N. J.; Uhlendorf, I. *J. Phys. Chem. B* **1997**, *101*, 10281.

(110) Franco, G.; Gehring, J.; Peter, L. M.; Ponomarev, E. A.; Uhlendorf, I. *J. Phys. Chem. B* **1999**, *103*, 692.

(111) Solbrand, A.; Henningsson, A.; Sodergren, S.; Lindstrom, H.; Hagfeldt, A.; Lindquist, S. E. *J. Phys. Chem. B* **1999**, *103*, 1078.

(112) Zaban, A.; Bisquert, J.; Chen, S.; Abayev, I.; Fabregat-Santiago, F. **2001**, Submitted for Publication.

(113) Sodergren, S.; Hagfeldt, A.; Olsson, J.; Lindquist, S. E. *J. Phys. Chem.* **1994**, *98*, 5552.

(114) Lyon, L. A.; Hupp, J. T. *J. Phys. Chem.* **1995**, *99*, 15718.

(115) Elliott, C. M.; Hershenhart, E. J. *J. Am. Chem. Soc.* **1982**, *104*, 7519.

(116) Nozik, A. J. *Ann. Rev. Phys. Chem.* **1978**, *29*, 189-222.

(117) Schlichthrol, G.; Huang, S. Y.; Sprague, J.; Frank, A. J. *J. Phys. Chem. B* **1997**, *101*, 8141.

(118) Fisher, A. C.; Peter, L. M.; Ponomarev, E. A.; Walker, A. B.; Wijayantha, K. G. U. *J. Phys. Chem. B* **2000**, *104*, 949.

INDEX

T

MIX
Papier aus verantwortungsvollen Quellen
Paper from responsible sources
FSC® C105338

If you have any concerns about our products,
you can contact us on
ProductSafety@springernature.com

In case Publisher is established outside the EU,
the EU authorized representative is:
Springer Nature Customer Service Center GmbH
Europaplatz 3, 69115 Heidelberg, Germany

Printed by Libri Plureos GmbH
in Hamburg, Germany